Miscellaneous Foods

Fourth supplement to the Fifth Edition of

McCance and Widdowson's

The Composition of Foods

Miscellaneous Foods

Fourth supplement to the Fifth Edition of

McCance and Widdowson's

The Composition of Foods

W. Chan, J. Brown and D. H. Buss

The Royal Society of Chemistry
and
Ministry of Agriculture, Fisheries and Food

A catalogue record for this book is available from the British Library.

ISBN 0-85186-360-4

Published by the The Royal Society of Chemistry, Cambridge, and the Ministry of Agriculture, Fisheries, and Food, London.

Photocomposed by Land and Unwin Ltd, Bugbrooke
Printed in the United Kingdom by the Bath Press, Bath

CONTENTS

ACKNOWLEDGEMENTS

Numerous people have helped during the preparation of this book.

Many new analyses were undertaken by the Laboratory of the Government Chemist, Rank Hovis McDougall and the Leatherhead Food Research Association. The analytical teams were headed by Mr I Lumley and Mrs G Holcombe, Dr M A Jordan and Dr P J Farnell respectively.

We wish to express our thanks to numerous manufacturers, retailers and other organisations for information on the range and composition of their products. In particular, we would like to thank Asda Stores Ltd, The Boots Company Ltd, Cow & Gate Nutricia Ltd, Dairy Crest Ltd, Farley Health Products Ltd, HJ Heinz Company Ltd, Marks and Spencer plc, Milupa Ltd, Robinsons Ltd, Safeway plc, Tesco Stores Ltd, and St Ivel Ltd.. KP foods, the Leatherhead Food Research Association and the National Dairy Council kindly provided the cover photographs.

We would also like to express our appreciation for all the help given to us by many people in the Ministry of Agriculture, Fisheries and Food (MAFF), the Royal Society of Chemistry (RSC) and elsewhere who were involved in the work leading up to the production of this book. In particular, we would like to thank Mr Peter Gillatt (MAFF) for his helpful advice, and Ms Bridie Holland (formerly RSC) for her invaluable contributions to the early stages of this supplement and for all her work on previous publications from the UK nutrient databank.

The final preparation of this book was overseen by a committee which, besides the authors, comprised Dr M C Edwards (Campden Food and Drink Research Association), Dr A M Fehily (HJ Heinz Company Ltd), Miss A A Paul (MRC Dunn Nutrition Centre), Mrs P M Richardson (Northwick Park Hospital), Professor D A T Southgate (formerly AFRC, Institute of Food Research), and Dr F J Taylor (RSC).

INTRODUCTION

This is the seventh detailed reference book on the nutrients in food, in a series replacing and extending the information in McCance and Widdowson's *The Composition of Foods*. It shows the nutrients in a wide range of foods including fats and oils; sugars, syrups and preserves; chocolate and sugar confectionery; savoury snacks; coffee and tea; soft drinks; alcoholic drinks, soups, sauces and pickles; baby foods; and a number of other miscellaneous foods and ingredients.

There has been a considerable increase in the number and variety of these foods in the United Kingdom in recent years, so this book includes information on 418 items – more than twice the number of such foods in the fifth edition of *The Composition of Foods* (Holland *et al.*, 1991b). There is also an increase in the number of nutrients shown. All the data for the foods included in the fifth edition have been thoroughly reviewed and most have been updated, while the data for the additional foods has been obtained from new analyses specifically for this book, or, where appropriate, from recent information from manufacturers.

These tables are part of a continuing series produced by the Royal Society of Chemistry (RSC) and the Ministry of Agriculture, Fisheries and Food (MAFF), who have been collaborating since 1987 on the development of a comprehensive and up-to-date database on nutrients in the wide range of foods now available in Britain. Other current supplements cover *Cereals and Cereal Products* (Holland *et al.*,1988), *Milk Products and Eggs* (Holland *et al.*,1989), *Vegetables, Herbs and Spices* (Holland *et al.*, 1991a), *Fruit and Nuts* (Holland *et al.*, 1992a), *Vegetable Dishes* (Holland *et al.*, 1992b), and *Fish and Fish Products* (Holland *et al.*, 1993). Computerised versions are also available, details of which can be obtained from the Royal Society of Chemistry.

Methods

The selection of foods and of nutrient values has followed the general principles used in the preparation of previous books in *The Composition of Foods* series. The foods are as far as possible those most widely available and nutritionally important in Britain at the present time, and the nutrient values are a mixture of direct analyses and appropriate values from the scientific literature and from manufacturers. Only the most recent analytical values have been included for those manufactured foods whose formulation and composition are changed from time to time. Many more values from manufacturers have been included in this book than in previous ones, partly because most of the foods in this supplement are manufactured foods but also because the range of foods now analysed by manufacturers has increased. There is comparatively little information on these items in the scientific literature, but some values have been included from food composition tables from other countries where the foods are known to be the same as in those countries.

Manufacturers' and literature values

Manufacturers' and literature information was only included where full details of the samples were known and where the samples were representative of the foods now available; where suitable methods of analysis had been used; and where the results were available in sufficient detail. There were nevertheless many gaps, and for these a large number of new analyses were commissioned.

Analyses

Where little or nothing was known of the nutrients in an important food, arrangements were made for its direct analysis in one of three laboratories. Detailed sampling and analytical protocols were devised for each item. Most of the foods were bought from a wide variety of shops in or near London. The samples were not normally analysed individually but, as for previous supplements, pooled according to market share before analysis.

The analytical methods for the major nutrients were as described in the fifth edition of *The Composition of Foods* (Holland *et al.*, 1991b), while those for the nutrients not included in that book are given in the supplements on *Milk Products and Eggs* or on *Vegetables, Herbs and Spices*. Individual fatty acids were determined as their methyl esters by capillary gas chromatography. Further details of each determination can be provided on request.

Arrangement of the tables

Food groups

For ease of reference, the values have been brought together in nine broad groups, and then further subdivided within those. Although the foods have been listed alphabetically within each section, the first set of values for each is generally for the product as bought, followed by values for the food after dilution or after any other preparation that may be needed before it can be eaten.

Numbering system

As in previous supplements, the foods have been numbered in sequence (from 1 to 418) with a unique two digit prefix. For this supplement, the prefix is '17'. The full code numbers for butteroil and dried sweetcorn, the first and last foods in this supplement, are thus 17-001 and 17-418, and these are the numbers that will be used in nutrient databank applications.

Description and number of samples

The information given under this heading includes examples of the products and selected trade names where this would be helpful, as well as the number and nature of samples taken for analysis. Some additional values for related foods were calculated from the analytical values, usually after the addition of water, and this is indicated under this heading, as are the major sources of values that were based on information from manufacturers and the scientific literature.

Nutrients

As in most previous books in this series, the nutrient values for each food are shown on four consecutive pages. The presentation of most of the nutrients follows the established pattern, but the information on the second page in the fats and oils section is different so that the information most appropriate to all the diverse types of food in this supplement can be covered in the four pages.

All nutrient values are given per 100 grams of the edible portion of the food, except for alcoholic beverages where the values are given per 100 ml.

Proximates: – The first page for each food shows the amounts of water, alcohol (for alcoholic beverages only), total nitrogen, protein, fat, and available carbohydrate expressed as its monosaccharide equivalent. These values are in grams, and then the foods energy value is given in kilocalories and in kilojoules. All values are per 100 g of edible matter, but in this supplement there is no indication of the proportion of edible matter in each food since it is always 1.00.

Protein was derived from the nitrogen values by multiplying them by 6.25 after subtracting any non-protein nitrogen, and the energy values were derived by multiplying the amounts of protein, fat, carbohydrate and (where appropriate) alcohol by the factors in **Table 1**. Where oligosaccharides, organic acids or polyols were present (for example the acetic acid in pickles, citric and malic acids in soft drinks and sorbitol in diabetic foods), their energy contribution was also included using the values in the Appendix on page 170. A few of the carbohydrate values from manufacturers and the literature have been estimated by subtracting the other proximates from 100 rather than by analysis; these and the corresponding energy values have been presented as quoted, but in italics to distinguish them from analytical values.

Table 1: – Energy conversion factors

	kcal/g	*kJ/g*
Protein	4	17
Fat	9	37
Available carbohydrate expressed as monosaccharide	3.75	16
Alcohol	7	29

Carbohydrates and 'fibre': – For most foods, the second page gives more details of the individual carbohydrates, fibre fractions and fat constituents. The value for total sugars is the sum of the glucose, fructose, sucrose, maltose and lactose in the food, but does not include the oligosaccharides that are present in significant quantities in a number of foods where glucose syrups or maltodextrins have been used. These oligosaccharides have, however, been included in the total carbohydrate on the preceding page wherever possible, and this value will then be greater than the sum of the starch and the sugars alone. As in the previous UK tables, the amounts of sugars, starch and available carbohydrate are shown after conversion to their monosaccharide equivalents, while all fibre values are the actual amounts of each component. The relationships between the various forms and fractions of fibre are shown in Table 2.

Table 2: – Relationships between the dietary fibre fractions

Component	Fibre fraction	NSP	Southgate
Cellulose	Insoluble fibre	Englyst fibre (non-starch polysaccharides)	Southgate fibre[a] (unavailable carbohydrate)
Insoluble non-cellulosic polysaccharides			
Soluble non-cellulosic polysaccharides	Soluble fibre		
'Lignin'			

[a] The Southgate values are generally higher than NSP values because they include substances measuring as lignin and also because the enzymic preparation used leaves some enzymatically resistant starch in the dietary fibre residue. A resistant starch value can be obtained from the NSP procedures, but because this uses different conditions and enzymes this may or may not be the same as the enzymatically resistant starch in the Southgate method.

Fats: - Values are given for the total saturated, monounsaturated and polyunsaturated fatty acids in each food. The unsaturated acids include both *cis* and *trans* isomers, but for the fats and oils there is an additional column showing the total amount of *trans* fatty acids in each food. Cholesterol is shown for all foods, but for the fats and oils the total amount of phytosterols (plant sterols) is also given. Further details of the individual fatty acids and phytosterols in selected foods are presented in the Appendixes on pages 154 and 166.

Minerals and vitamins: - The range of minerals and vitamins shown is the same as in previous books. The values for total carotene and for vitamin E have been corrected for the relative activities of the different fractions using the factors given in the fifth edition of *The Composition of Foods* (Holland *et al.*, 1991b). An Appendix on page 163 shows further details of the individual vitamin E fractions where they are known.

Appendices

This supplement contains a number of appendices. The first gives the individual fatty acids in selected fats and oils, and is followed by appendices showing the amounts of vitamin E fractions and the main phytosterols in selected foods. Further appendices give the proximates and energy value for a number of miscellaneous food ingredients and additives, and the percentage of alcohol (and typical ranges) *by volume* in different beers. These are followed by details of the recipes used in this supplement.

Nutrient variability

Almost all foods vary somewhat in composition, and this is equally true for most of the manufactured foods included in this supplement. This is partly because there will be natural variations in the composition and proportions of their ingredients, but it is also important to remember that the ingredients used, or the proportions of the ingredients, in many of them can suddenly be substantially altered with no change in the product name or description. This may reflect changing raw material prices, but there can also be changes in, for example, the amounts of salt, sugar, fats, water and micronutrients added in order to meet changing health considerations. The proportions of the ingredients, and therefore the nutritional value, may also change if the product size is changed. It should further be borne in mind that the values for a number of foods in this book are averages based on a number of similar foods, but the nutritional value of any one of the component samples may be significantly different from this average. It therefore remains important when using the values in these tables to ensure that the product is as similar as possible to that described here.

A further point to bear in mind is that for some related foods, including a number of the beverages, snacks and baby foods in these tables, the apparent (but usually small) differences in composition may reflect analytical variations as much as real differences in composition.

The introduction to the fifth edition of *The Composition of Foods* contains a more detailed description of these and many other factors that should be taken into account in the proper use of food composition tables. Users of the present supplement are advised to read them and take them to heart.

References to introductory text

Holland, B., Unwin, I.D., and Buss, D.H. (1988) *Cereals and Cereal Products.* Third supplement to *McCance and Widdowson's The Composition of Foods*, Royal Society of Chemistry, Cambridge

Holland, B., Unwin, I.D., and Buss, D.H. (1989) *Milk Products and Eggs.* Fourth supplement to *McCance and Widdowson's The Composition of Foods*, Royal Society of Chemistry, Cambridge

Holland, B., Unwin, I.D., and Buss, D.H. (1991a) *Vegetables, Herbs and Spices.* Fifth supplement to *McCance and Widdowson's The Composition of Foods*, Royal Society of Chemistry, Cambridge

Holland, B., Welch, A.A., Unwin, I.D., Buss, D.H., Paul, A.A. and Southgate, D.A.T. (1991b) *McCance and Widdowson's The Composition of Foods*, 5th edition, Royal Society of Chemistry, Cambridge

Holland, B., Unwin, I.D., and Buss, D.H. (1992a) *Fruit and Nuts.* First supplement to 5th edition of *McCance and Widdowson's The Composition of Foods.* Royal Society of Chemistry, Cambridge

Holland, B., Welch, A.A., and Buss, D.H. (1992b) *Vegetable Dishes.* Second supplement to 5th edition of *McCance and Widdowson's The Composition of Foods.* Royal Society of Chemistry, Cambridge

Holland, B., Brown, J., and Buss, D.H. (1993) *Fish and Fish products.* Third supplement to 5th edition of *McCance and Widdowson's The Composition of Foods.* Royal Society of Chemistry, Cambridge

The Tables

Symbols and abbreviations used in the tables

Symbols

0	None of the nutrient is present
Tr	Trace
N	The nutrient is present in significant quantities but there is no reliable information on the amount
( )	Estimated value, or water values estimated by difference
Italic text	Carbohydrate estimated 'by difference', and energy values based upon these quantities

Abbreviations

Gluc	Glucose
Fruct	Fructose
Sucr	Sucrose
Malt	Maltose
Lact	Lactose
Satd	Saturated
Monounsatd	Monounsaturated
Polyunsatd	Polyunsaturated
Trypt	Tryptophan
equiv	equivalents

FATS AND OILS

This section includes entries for cooking fats, butter, different types of margarines, blends which contain mixtures of butter and vegetable fats, fat spreads based on vegetable oils only, and oils. Most of the values in the fifth edition of *The Composition of Foods* have been updated and many new fats have been included.

The layout of page 2 is slightly different from that in the other sections and in addition to the total amounts of saturated, monounsaturated and polyunsaturated fatty acids (which include any trans fatty acids) there is also a value for total trans fatty acids in each product. Cholesterol and total phytosterols are also shown, but there are no values for individual sugars or fibre. The amounts of individual phytosterols in selected margarines, blends, fat spreads and oils are given on page 166.

Manufacturers frequently vary the blend of oils and fats in cooking and spreading fats, which will alter the proportion of fatty acids. Users requiring fatty acids data for specific products may wish to contact the manufacturers directly.

The fatty acid profile for blended vegetable oil was calculated from the values for the main components (soya, rape and corn oils). Although the proportions will vary, this entry has been included for recipe calculation purposes and for dietary survey work where unspecified vegetable oil has been used or consumed.

More detailed fatty acid profiles of selected fats and oils are given in the Appendix on page 154. The fatty acid composition of oils represent those of crude oils, and should not change significantly during the refining process.

British regulations require margarines to contain the equivalent of 800–1000µg vitamin A and 7–9µg vitamin D per 100g. Similar additions are made to many but not all reduced fat spreads and vitamin E may also be added. The values in these tables will reflect the proportion of the products within each category which are fortified. The amounts in specific brands may be obtained from the label or the manufacturer.

Composition of food per 100g

No. 17-	Food	Description and main data sources	Water g	Total Nitrogen g	Protein g	Fat g	Carbo-hydrate g	Energy value kcal	Energy value kJ
	Cooking fats								
1	**Butteroil**, unsalted	Calculated from butter	0.5	0.10	0.6	99.3	Tr	896	3684
2	**Cocoa butter**	Analysis and literature sources	(0.5)	Tr	Tr	99.5	0	896	3682
3	**Cocoa butter alternative**	Analysis and literature sources; mixture of cocoa butter equivalent, replacer and substitute	(0.3)	Tr	Tr	99.7	0	897	3689
4	**Compound cooking fat**	10 samples of a mixture of Cookeen and White Cap	Tr	Tr	Tr	99.9	0	899	3696
5	polyunsaturated	10 samples of White Flora	Tr	Tr	Tr	99.9	Tr	899	3696
6	**Dripping**, beef	Analysed as purchased	1.0	Tr	Tr	99.0	Tr	891	3663
7	**Ghee**, butter	5 assorted samples	0.1	Tr	Tr	99.8	Tr	898	3693
8	palm	5 samples of the same brand	0.1	Tr	Tr	99.7	Tr	897	3689
9	vegetable	5 samples; different types	0.1	Tr	Tr	99.4	Tr	895	3678
10	**Lard**	6 samples; 3 brands	1.0	Tr	Tr	99.0	0	891	3663
11	**Suet**, shredded	6 samples of the same brand	1.5	Tr	Tr	86.7	12.1	826	3402
12	vegetable	10 samples; 5 brands	0.8	0.19	1.2	87.9	*10.1*	*836*	*3444*
	Spreading fats								
13	**Butter**	Analysis and literature sources	15.6[a]	0.08	0.5	81.7[a]	Tr	737	3031
14	spreadable	8 samples; different brands	15.5	0.08	0.5	82.5	Tr	745	3061
15	**Blended spread** (70-80% fat)	30 samples including Clover, Golden Crown and Willow	21.0	0.10	0.6	74.8	1.1	680	2795
16	(40% fat)	20 samples including Anchor half fat butter and Clover Extra Light	51.4	1.02	6.5	40.3	0.4	390	1608

[a] Unsalted butter contains 15.7g water and 82.7g fat per 100g

Carbohydrates and fatty acids, g per 100g food
Cholesterol and phytosterols, mg per 100 g food

No. 17-	Food	Starch g	Total sugars g	Fatty acids: Satd g	Fatty acids: Mono-unsatd g	Fatty acids: Poly-unsatd g	Fatty acids: Trans g	Cholesterol mg	Total phytosterols mg
	Cooking fats								
1	**Butteroil**, unsalted	0	Tr	65.6	24.1	3.2	(4.1)	280	(5)
2	**Cocoa butter**	0	0	59.0	32.8	3.3	N	3	221
3	**Cocoa butter alternative**	0	Tr	59.4	32.5	2.8	N	0	357
4	**Compound cooking fat**	0	0	49.5	41.2	5.3	16.4	425	58
5	polyunsaturated	0	Tr	20.5	31.1	43.8	7.4	10	34
6	**Dripping**, beef	0	Tr	52.5	38.9	2.5	(4.6)	94	N
7	**Ghee**, butter	0	Tr	66.0	24.1	3.4	N	280	N
8	palm	0	Tr	47.0	35.3	8.9	N	0	N
9	vegetable	0	Tr	48.4	37.0	9.7	1.1	0	51
10	**Lard**	0	0	40.3	43.4	10.0	Tr	93	N
11	**Suet**, shredded[a]	11.9	0.2	49.9	30.4	2.2	(4.0)	82	N
12	vegetable	10.1	0	45.0	26.3	12.8	21.8	0	N
	Spreading fats								
13	**Butter**	0	Tr	54.8	20.0	2.7	3.4	230	4
14	spreadable	0	Tr	51.2	26.0	4.5	2.4	280	Tr
15	**Blended spread** (70-80% fat)	0	1.1	25.5	37.5	8.5	9.5	67	268
16	(40% fat)	0	0.4	18.1	13.4	7.3	4.9	46	63

[a] Shredded suet contains 0.6g Southgate fibre and 0.5g Englyst fibre per 100g

Inorganic constituents per 100g food

No. 17-	Food	mg										µg	
		Na	K	Ca	Mg	P	Fe	Cu	Zn	Cl	Mn	Se	I
	Cooking fats												
1	**Butteroil**, unsalted	15	18	18	2	29	0.2	0.04	0.1	22	Tr	Tr	46
2	**Cocoa butter**	Tr	Tr	Tr	Tr	Tr	Tr	Tr	Tr	Tr	Tr	Tr	N
3	**Cocoa butter alternative**	Tr	Tr	Tr	Tr	Tr	Tr	Tr	Tr	Tr	Tr	Tr	N
4	**Compound cooking fat**	Tr	Tr	Tr	Tr	Tr	Tr	Tr	Tr	Tr	Tr	Tr	Tr
5	polyunsaturated	Tr	Tr	Tr	Tr	Tr	Tr	Tr	Tr	Tr	Tr	Tr	Tr
6	**Dripping**, beef	5	4	1	Tr	13	0.2	N	N	2	Tr	Tr	(5)
7	**Ghee**, butter	2	3	Tr	Tr	Tr	0.2	Tr	Tr	28	Tr	Tr	N
8	palm	1	1	Tr	Tr	Tr	0.1	0.21	Tr	N	Tr	Tr	Tr
9	vegetable	1	1	Tr	Tr	Tr	Tr	0.14	Tr	N	Tr	N	N
10	**Lard**	2	1	1	1	3	0.1	0.02	N	4	Tr	Tr	Tr
11	**Suet**, shredded	Tr	Tr	Tr	Tr	Tr	Tr	Tr	Tr	Tr	Tr	N	5
12	vegetable	10	Tr	Tr	Tr	Tr	Tr	Tr	Tr	N	Tr	Tr	Tr
	Spreading fats												
13	**Butter**	750[a]	15	15	2	24	0.2	0.03	0.1	1150[a]	Tr	Tr	38
14	spreadable	390[b]	(15)	(15)	(2)	(24)	(0.2)	(0.03)	(0.1)	640	Tr	Tr	(38)
15	**Blended spread** (70-80% fat)	670	43	14	2	18	Tr	Tr	Tr	1010	Tr	N	N
16	(40% fat)	510	N	N	N	N	N	N	N	780	Tr	N	N

[a] Unsalted butter contains 11mg Na and 17mg Cl per 100g

[b] Average of salted and unsalted. Salted versions contain between 800 to 1500mg Na per 100g

No. 17-	Food	Retinol µg	Carotene µg	Vitamin D µg	Vitamin E mg	Thiamin mg	Riboflavin mg	Niacin mg	Trypt/60 mg	Vitamin B_6 mg	Vitamin B_{12} µg	Folate µg	Pantothenate mg	Biotin µg	Vitamin C mg
	Cooking fats														
1	**Butteroil**, unsalted	990	525	0.9	2.4	Tr	(0.02)	Tr	0.1	Tr	Tr	Tr	(0.05)	Tr	0
2	**Cocoa butter**	N	N	Tr	1.1	Tr	Tr	Tr	Tr	Tr	Tr	Tr	Tr	Tr	0
3	**Cocoa butter alternative**	0	Tr	0	N	Tr	Tr	Tr	Tr	Tr	0	Tr	Tr	Tr	0
4	**Compound cooking fat**	0	0	0	Tr	0	0	0	0	0	0	0	0	0	0
5	polyunsaturated	0	0	0	Tr	0	0	0	0	0	0	0	0	0	0
6	**Dripping**, beef	N	N	Tr	(0.3)	Tr	Tr	Tr	Tr	Tr	Tr	Tr	Tr	Tr	0
7	**Ghee**, butter	675	500	1.9	3.3	0	Tr	Tr	Tr	Tr	Tr	0	Tr	Tr	0
8	palm	Tr	Tr	0	7.4	0	0	Tr	Tr	Tr	0	0	Tr	Tr	0
9	vegetable	Tr	Tr	0	10.3[a]	0	0	Tr	Tr	Tr	0	0	Tr	Tr	0
10	**Lard**	Tr	0	N	1.0	Tr	Tr	Tr	Tr	Tr	Tr	Tr	Tr	Tr	0
11	**Suet**, shredded	52	73	Tr	1.5	Tr	Tr	Tr	Tr	Tr	Tr	Tr	Tr	Tr	0
12	vegetable	0	0	0	18.0	Tr	Tr	Tr	Tr	Tr	0	Tr	Tr	Tr	0
	Spreading fats														
13	**Butter**	815	430	0.8	2.0	Tr	(0.02)	Tr	0.1	Tr	Tr	Tr	(0.04)	Tr	0
14	spreadable	(815)	670	Tr	2.9	Tr	(0.02)	Tr	(0.1)	Tr	Tr	Tr	(0.04)	Tr	0
15	**Blended spread** (70-80% fat)	565	445	4.1	11.3[a]	Tr	Tr	Tr	0.1	Tr	Tr	Tr	Tr	Tr	0
16	(40% fat)	160	430	0.2	3.9[a]	Tr	Tr	Tr	Tr	Tr	Tr	Tr	Tr	Tr	0

[a] The vitamin E content will vary according to the type of oil

No. 17-	Food	Description and main data sources	Water g	Total Nitrogen g	Protein g	Fat g	Carbo-hydrate g	Energy value kcal	Energy value kJ
	Spreading fats *continued*								
17	**Dairy spread** (40% fat)	Manufacturers' data on own brands	(52.9)	1.10	7.0	40.0	0.1	388	1601
18	**Margarine**, hard, animal and vegetable fats	10 samples of Echo and Stork	16.0	0.03	0.2	79.3	1.0	718	2954
19	-, vegetable fats only	4 samples of Tomor	16.0	0.03	0.2	81.6	1.0	739	3039
20	soft, not polyunsaturated	20 samples of a mixture of Stork SB and own brands soft margarine	16.0	0.03	0.2	81.7	1.0	740	3042
21	-, polyunsaturated	20 samples of a mixture of Blue Band and own brands soya margarine	16.0	Tr	Tr	82.8	0.2	746	3067
22	**Fat spread** (70-80% fat), not polyunsaturated	10 samples including Krona Gold	22.0	0.06	0.4	71.2	Tr	642	2641
23	(70% fat), polyunsaturated	10 samples including I Can't Believe it's Not Butter and Flora	(26.6)	0.08	0.5	70.0	0.8	635	2611
24	(60% fat), polyunsaturated	10 samples including Vitalite Light	(37.7)	0.03	0.2	60.8	1.3	553	2274
25	-, with olive oil	5 samples including Olivio and own brands	38.2	0.02	0.1	62.7	1.1	569	2339
26	(40% fat), not polyunsaturated	20 samples including Gold and Delight	52.3	1.02	6.5	40.2	1.3	393	1619
27	(35-40% fat), polyunsaturated	Manufacturers' data (Gold Sunflower and Flora Extra Light)	53.8	0.76	4.9	37.6	1.8	365	1503
28	(20-25% fat), not polyunsaturated	20 samples including Gold Lowest and Outline	59.0	0.92	5.9	23.8	2.5	247	1021
29	-, polyunsaturated	Manufacturers' data on own brands	(78.7)	0	0	20.0	0.8	183	753
30	(5% fat)	Manufacturers' data on own brands	(80.0)	0.71	4.5	5.0	10.5	102	430

Carbohydrates and fatty acids, g per 100g food
Cholesterol and phytosterols, mg per 100 g food

No. 17-	Food	Starch g	Total sugars g	Fatty acids: Satd g	Fatty acids: Mono-unsatd g	Fatty acids: Poly-unsatd g	Fatty acids: Trans g	Cholesterol mg	Total phytosterols mg
	***Spreading fats** continued*								
17	**Dairy spread** (40% fat)	0	0.1	26.8	10.0	1.2	N	N	N
18	**Margarine**, hard, animal and vegetable fats	0	1.0	34.6	36.2	5.4	12.2	285	68
19	-, vegetable fats only	0	1.0	35.9	33.0	9.3	14.6	15	220
20	soft, not polyunsaturated	0	1.0	27.2	38.9	12.4	9.8	275	187
21	-, polyunsaturated	0	0.2	17.0	26.6	36.0	6.7	2	242
22	**Fat spread** (70-80% fat), not polyunsaturated	0	Tr	30.4	31.2	6.5	2.0	86	43
23	(70% fat), polyunsaturated	0	0.8	12.6	21.2	33.1	3.0	Tr	262
24	(60% fat), polyunsaturated	0	1.3	11.3	18.1	28.6	3.3	3	197
25	-, with olive oil	0	1.1	11.3	36.4	12.5	6.0	0	223
26	(40% fat), not polyunsaturated	0	1.3	11.7	20.6	6.4	4.1	6	181
27	(35-40% fat), polyunsaturated	0.6	1.3	8.9	9.4	18.0	0.7	Tr	N
28	(20-25% fat), not polyunsaturated	Tr	1.1	6.2	11.9	4.8	2.3	8	113
29	-, polyunsaturated	0.8	0	3.7	7.2	9.1	N	Tr	N
30	(5% fat)	0	3.0	0.9	2.3	1.9	N	N	N

No. 17-	Food	mg										µg	
		Na	K	Ca	Mg	P	Fe	Cu	Zn	Cl	Mn	Se	I
	***Spreading fats** continued*												
17	**Dairy spread** (40% fat)	600	N	N	N	N	N	N	N	N	N	N	N
18	**Margarine**, hard, animal and vegetable fats	940	5	4	1	12	0.3	0.04	N	1200	Tr	Tr	N
19	-, vegetable fats only	800	5	4	1	12	0.3	0.04	N	1200	Tr	Tr	N
20	soft, not polyunsaturated	880	5	4	1	12	0.3	0.04	N	(1320)	Tr	Tr	N
21	-, polyunsaturated	680	5	4	1	12	0.3	0.04	N	(1020)	Tr	Tr	N
22	**Fat spread** (70-80% fat), not polyunsaturated	1060	43	14	2	18	Tr	Tr	Tr	1270	Tr	N	N
23	(70% fat), polyunsaturated	800	N	N	N	N	Tr	Tr	N	1200	Tr	N	N
24	(60% fat), polyunsaturated	710	N	N	N	N	Tr	Tr	N	1070	Tr	N	N
25	-, with olive oil	600	N	N	N	N	Tr	Tr	N	910	Tr	N	N
26	(40% fat), not polyunsaturated	650	110	39	4	82	Tr	Tr	0.2	800	Tr	N	N
27	(35-40% fat), polyunsaturated	650	N	N	N	N	Tr	Tr	N	990	Tr	N	N
28	(20-25% fat), not polyunsaturated	540	630	N	N	N	Tr	Tr	N	(830)	Tr	N	N
29	-, polyunsaturated	500	N	N	N	N	Tr	Tr	N	(770)	Tr	N	N
30	(5% fat)	500	N	N	N	N	Tr	Tr	N	760	Tr	N	N

No. 17-	Food	Retinol µg	Carotene[a] µg	Vitamin D µg	Vitamin[b] E mg	Thiamin mg	Ribo-flavin mg	Niacin mg	Trypt/60 mg	Vitamin B_6 mg	Vitamin B_{12} µg	Folate µg	Panto-thenate mg	Biotin µg	Vitamin C mg
	***Spreading fats** continued*														
17	**Dairy spread**, (40% fat)	N	N	N	N	Tr	Tr	Tr	Tr	Tr	Tr	Tr	Tr	Tr	0
18	**Margarine**, hard, animal and vegetable fats	(665)	(750)	(7.9)	4.4	Tr	Tr	Tr	Tr	Tr	Tr	Tr	Tr	Tr	0
19	-, vegetable fats only	(665)	(750)	(7.9)	N	Tr	Tr	Tr	Tr	Tr	0	Tr	Tr	Tr	0
20	soft, not polyunsaturated	745	445	7.8	12.3	Tr	Tr	Tr	Tr	Tr	Tr	Tr	Tr	Tr	0
21	-, polyunsaturated	675	350	(7.9)	32.6	Tr	Tr	Tr	Tr	Tr	Tr	Tr	Tr	Tr	0
22	**Fat spread** (70-80% fat), not polyunsaturated	40	330	5.8	2.5	Tr	Tr	Tr	0.1	Tr	Tr	Tr	Tr	Tr	0
23	(70% fat), polyunsaturated	N	N	N	38.0	Tr	Tr	Tr	Tr	Tr	Tr	Tr	Tr	Tr	0
24	(60% fat), polyunsaturated	980	Tr	N	30.8	Tr	Tr	Tr	Tr	Tr	Tr	Tr	Tr	Tr	0
25	-, with olive oil	N	N	N	N	Tr	Tr	Tr	Tr	Tr	0	Tr	Tr	Tr	0
26	(40% fat), not polyunsaturated	650	805	8.4	8.0	Tr	Tr	Tr	1.4	Tr	Tr	Tr	Tr	Tr	0
27	(35-40% fat), polyunsaturated	N	N	N	N	Tr	Tr	Tr	Tr	Tr	Tr	Tr	Tr	Tr	0
28	(20-25% fat), not polyunsaturated	470	575	7.8	5.1	Tr	Tr	Tr	1.9	Tr	Tr	Tr	Tr	Tr	0
29	-, polyunsaturated	N	N	N	N	Tr	Tr	Tr	Tr	Tr	Tr	Tr	Tr	Tr	0
30	(5% fat)	N	N	N	N	Tr	Tr	Tr	Tr	Tr	Tr	Tr	Tr	Tr	0

[a] Some brands may not contain β-carotene

[b] The vitamin E content will vary according to the blend of oils used

Composition of food per 100g

No. 17-	Food	Description and main data sources	Water g	Total Nitrogen g	Protein g	Fat g	Carbo-hydrate g	Energy value kcal	Energy value kJ
	Oils								
31	**Coconut oil**	Mean of 35 samples	Tr	Tr	Tr	99.9	0	899	3696
32	**Cod liver oil**	Mean of 20 samples	Tr	Tr	Tr	99.9	0	899	3696
33	**Corn oil**	Mean of 42 samples	Tr	Tr	Tr	99.9	0	899	3696
34	**Cottonseed oil**	Mean of 55 samples	Tr	Tr	Tr	99.0	0	891	3663
35	**Evening primrose oil**	Mean of 35 samples	Tr	Tr	Tr	99.9	0	899	3696
36	**Grapeseed oil**	Mean of 23 samples	Tr	Tr	Tr	99.9	0	899	3696
37	**Hazelnut oil**	Mean of 10 samples	Tr	Tr	Tr	99.9	0	899	3696
38	**Olive oil**	Mean of 35 samples; including virgin and extra virgin olive oil	Tr	Tr	Tr	99.9	0	899	3696
39	**Palm oil**	Mean of 55 samples	Tr	Tr	Tr	99.9	0	899	3696
40	**Peanut oil**	Mean of 71 samples	Tr	Tr	Tr	99.9	0	899	3696
41	**Rapeseed oil**	Mean of 100 samples	Tr	Tr	Tr	99.9	0	899	3696
42	**Safflower oil**	Mean of 28 samples	Tr	Tr	Tr	99.9	0	899	3696
43	**Sesame oil**	Mean of 22 samples and literature sources	0.1	0.03	0.2	99.7	0	898	3692
44	**Soya oil**	Mean of 39 samples	Tr	Tr	Tr	99.9	0	899	3696
45	**Sunflower oil**	Mean of 46 samples	Tr	Tr	Tr	99.9	0	899	3696
46	**Vegetable oil**, blended, average	Unspecified	Tr	Tr	Tr	99.9	0	899	8696
47	**Walnut oil**	Mean of 13 samples	Tr	Tr	Tr	99.9	0	899	3696
48	**Wheatgerm oil**	Mean of 35 samples	Tr	Tr	Tr	99.9	0	899	3696

Carbohydrates and fatty acids, g per 100g food
Cholesterol and phytosterols, mg per 100 g food

No. 17-	Food	Starch g	Total sugars g	Fatty acids Satd g	Fatty acids Mono-unsatd g	Fatty acids Poly-unsatd g	Fatty acids Trans g	Cholesterol mg	Total phytosterols mg
	Oils								
31	**Coconut oil**	0	0	86.5	6.0	1.5	Tr	0	88
32	**Cod liver oil**	0	0	20.2	44.6	29.9	Tr	(570)	Tr
33	**Corn oil**	0	0	14.4	29.9	51.3	Tr	0	845
34	**Cottonseed oil**	0	0	26.1	18.3	50.2	Tr	0	328
35	**Evening primrose oil**	0	0	7.8	10.6	76.6	Tr	0	N
36	**Grapeseed oil**	0	0	11.1	15.8	68.2	Tr	0	126
37	**Hazelnut oil**	0	0	7.8	76.5	11.2	Tr	0	N
38	**Olive oil**	0	0	14.3	73.0	8.2	0	0	114
39	**Palm oil**	0	0	47.8	37.1	10.4	Tr	0	39
40	**Peanut oil**	0	0	20.0	44.4	31.0	Tr	0	234
41	**Rapeseed oil**	0	0	6.6	59.3	29.3	Tr	0	251
42	**Safflower oil**	0	0	9.7	12.0	74.0	Tr	0	325
43	**Sesame oil**	0	0	14.6	37.5	43.4	Tr	0	654
44	**Soya oil**	0	0	15.6	21.3	58.8	Tr	0	330
45	**Sunflower oil**	0	0	12.0	20.5	63.3	Tr	0	290
46	**Vegetable oil**, blended, average	0	0	10.4[a]	35.5[a]	48.2[a]	Tr	0	N
47	**Walnut oil**	0	0	9.1	16.5	69.9	Tr	0	164
48	**Wheatgerm oil**	0	0	18.6	16.6	60.4	Tr	0	525

[a] The fatty acid profile will depend on the blend of oil used

Inorganic constituents per 100g food

No. 17-	Food	mg										µg	
		Na	K	Ca	Mg	P	Fe	Cu	Zn	Cl	Mn	Se	I
	Oils												
31	**Coconut oil**	Tr	Tr	Tr	Tr	Tr	Tr	Tr	Tr	Tr	Tr	Tr	Tr
32	**Cod liver oil**	Tr	Tr	Tr	Tr	Tr	Tr	Tr	Tr	Tr	Tr	Tr	Tr
33	**Corn oil**	Tr	Tr	Tr	Tr	Tr	0.1	0.01	Tr	Tr	Tr	Tr	Tr
34	**Cottonseed oil**	Tr	Tr	Tr	Tr	Tr	Tr	Tr	Tr	Tr	Tr	Tr	Tr
35	**Evening primrose oil**	Tr	Tr	Tr	Tr	Tr	Tr	Tr	Tr	Tr	Tr	Tr	Tr
36	**Grapeseed oil**	Tr	Tr	Tr	Tr	Tr	Tr	Tr	Tr	Tr	Tr	Tr	Tr
37	**Hazelnut oil**	Tr	Tr	Tr	Tr	Tr	Tr	Tr	Tr	Tr	Tr	Tr	Tr
38	**Olive oil**	Tr	Tr	Tr	Tr	Tr	0.4	0.01	Tr	Tr	Tr	Tr	Tr
39	**Palm oil**	Tr	Tr	Tr	Tr	Tr	0.4	Tr	Tr	Tr	Tr	Tr	Tr
40	**Peanut oil**	Tr	Tr	Tr	Tr	Tr	Tr	Tr	Tr	Tr	Tr	Tr	Tr
41	**Rapeseed oil**	Tr	Tr	Tr	Tr	Tr	0.1	0.01	Tr	Tr	Tr	Tr	Tr
42	**Safflower oil**	Tr	Tr	Tr	Tr	Tr	Tr	Tr	Tr	Tr	Tr	Tr	Tr
43	**Sesame oil**	2	20	10	Tr	N	0.1	Tr	Tr	N	Tr	Tr	Tr
44	**Soya oil**	Tr	Tr	Tr	Tr	Tr	0.1	0.01	Tr	Tr	Tr	Tr	Tr
45	**Sunflower oil**	Tr	Tr	Tr	Tr	Tr	0.1	0.01	Tr	Tr	Tr	Tr	Tr
46	**Vegetable oil**, blended, average	Tr	Tr	Tr	Tr	Tr	Tr	Tr	Tr	Tr	Tr	Tr	11
47	**Walnut oil**	Tr	Tr	Tr	Tr	Tr	Tr	Tr	Tr	Tr	Tr	Tr	Tr
48	**Wheatgerm oil**	Tr	Tr	Tr	Tr	Tr	Tr	Tr	Tr	Tr	Tr	Tr	Tr

No. 17-	Food	Retinol µg	Carotene µg	Vitamin D µg	Vitamin E mg	Thiamin mg	Ribo-flavin mg	Niacin mg	Trypt 60 mg	Vitamin B_6 mg	Vitamin B_{12} µg	Folate µg	Panto-thenate mg	Biotin µg	Vitamin C mg
Oils															
31	**Coconut oi**	0	Tr	0	0.7	Tr	Tr	Tr	Tr	Tr	0	Tr	Tr	Tr	0
32	**Cod liver oil**	1800	Tr	210.0	20.0	Tr	Tr	Tr	Tr	Tr	Tr	Tr	Tr	Tr	0
33	**Corn oil**	0	Tr	0	17.2	Tr	Tr	Tr	Tr	Tr	0	Tr	Tr	Tr	0
34	**Cottonseed oil**	0	Tr	0	42.8	Tr	Tr	Tr	Tr	Tr	0	Tr	Tr	Tr	0
35	**Evening primrose oil**	0	Tr	0	N	Tr	Tr	Tr	Tr	Tr	0	Tr	Tr	Tr	0
36	**Grapeseed oil**	0	Tr	0	N	Tr	Tr	Tr	Tr	Tr	0	Tr	Tr	Tr	0
37	**Hazelnut oil**	0	Tr	0	N	Tr	Tr	Tr	Tr	Tr	0	Tr	Tr	Tr	0
38	**Olive oil**	0	N	0	5.1	Tr	Tr	Tr	Tr	Tr	0	Tr	Tr	Tr	0
39	**Palm oil**	0	Tr[a]	0	33.1	Tr	Tr	Tr	Tr	Tr	0	Tr	Tr	Tr	0
40	**Peanut oil**	0	Tr	0	15.2	Tr	Tr	Tr	Tr	Tr	0	Tr	Tr	Tr	0
41	**Rapeseed oil**	0	Tr	0	22.2	Tr	Tr	Tr	Tr	Tr	0	Tr	Tr	Tr	0
42	**Safflower oil**	0	Tr	0	40.7	Tr	Tr	Tr	Tr	Tr	0	Tr	Tr	Tr	0
43	**Sesame oil**	0	Tr	0	N	0.01	0.07	0.1	Tr	Tr	0	Tr	Tr	Tr	0
44	**Soya oil**	0	Tr	0	16.1	Tr	Tr	Tr	Tr	Tr	0	Tr	Tr	Tr	0
45	**Sunflower oil**	0	Tr	0	49.2	Tr	Tr	Tr	Tr	Tr	0	Tr	Tr	Tr	0
46	**Vegetable oil**, blended, average	0	Tr	0	N[b]	Tr	Tr	Tr	Tr	Tr	0	Tr	Tr	Tr	0
47	**Walnut oil**	0	Tr	0	N	Tr	Tr	Tr	Tr	Tr	0	Tr	Tr	Tr	0
48	**Wheatgerm oil**	0	Tr	0	136.7	Tr	Tr	Tr	Tr	Tr	0	Tr	Tr	Tr	0

[a] Unrefined palm oil contains approximately 30000µg β- and 24000µg α- carotene per 100g
[b] The vitamin E content will vary according to the type of oil

SUGARS, PRESERVES AND CONFECTIONERY

This section includes data for sugars, syrups, preserves and chocolate and sugar confectionery. Most of the values in the fifth edition of *The Composition of Foods* have been updated and many new products have been included. This section of the book now contains values for 73 foods compared with 34 in the fifth edition.

The composition of many confectionery items may change if the product size or the recipe has changed. Amounts of protein, fat, carbohydrate and energy for many products are given on the label. Users are advised to check the label for more specific information.

Because the amounts of carbohydrate are given in these tables as their monosaccharide equivalent, they sometimes exceed 100g per 100g product. To convert the amounts of sucrose, maltose and lactose to the actual weight of sugar, divide by 1.05.

No. 17-	Food	Description and main data sources	Water g	Total Nitrogen g	Protein g	Fat g	Carbo-hydrate g	Energy value kcal	Energy value kJ
	Sugars and syrups								
49	**Glucose liquid**, BP	1 sample	20.4	Tr	Tr	0	84.7[a]	318	1355
50	**Honey**	8 samples; assorted types	17.5	0.06	0.4	0	75.8	288	1229
51	**Honeycomb**	2 samples, honey and comb together	20.2	0.09	0.6	4.6[b]	74.4	281	1201
52	**Ice Magic sauce**	9 samples of the same brand; assorted chocolate flavours	0.5	0.80	5.0	38.0	31.5	480	1995
53	**Ice cream sauce**, topping	8 samples, 3 brands; strawberry and chocolate flavours	40.6	0.13	0.8	0.2	53.9	207	883
54	**Icing**, butter	Recipe	8.7	0.03	0.2	29.2	63.9	507[c]	2122[c]
55	fondant	Recipe	12.1	0.16	1.0	Tr	90.6	344	1467
56	glacé	Recipe	15.3	Tr	Tr	0	89.9	337	1439
57	Royal	Recipe	9.2	0.14	0.9	Tr	95.1	360	1536
58	**Jaggery**	5 assorted samples	3.4	0.08	0.5	0	97.2	367	1564
59	**Molasses**	Ref. Cutrufelli and Pehrsson (1991)	25.9	0	0	0.1	*68.8*	*266*	*1115*
60	**Sugar**, brown	3 samples	Tr	0.02	0.1	0	101.3	362[d]	1546[d]
61	Demerara	5 samples	Tr	0.08	0.5	0	104.5	394[d]	1681[d]
62	icing	Analysis and maunfacturers' data	Tr	Tr	Tr	0	104.9	393	1678
63	white	Granulated and loaf sugar	Tr	Tr	Tr	0	105.0	394	1680
64	**Syrup**, corn, dark	Ref. Cutrufelli and Pehrsson (1991)	22.8	0	0	0	*76.6*[e]	*282*	*1181*
65	golden	3 samples of the same brand	20.0	0.05	0.3	0	79.0	298	1269
66	-, pouring	2 samples of the same brand	18.1	Tr	Tr	0	79.0	296	1264
67	maple	Ref. Cutrufelli and Pehrsson (1991)	32.0	0	0	0.2	*67.2*	*262*	*1095*
68	**Treacle**, black	3 samples	28.5	0.19	1.2	0	67.2	257	1096

[a] Includes oligosaccharides
[b] Waxy material, probably not available as fat; disregarded in calculating energy values
[c] Includes contribution from 0.5g alcohol
[d] Light muscovado sugar provides 376kcal, 1705kJ per 100g. Dark muscovado sugar provides 355kcal, 1607kJ per 100g
[e] Includes 8.9g oligosaccharides

Carbohydrate fractions and fatty acids, g per 100g food
Cholesterol, mg per 100g food

No. 17-	Food	Starch g	Total sugars g	Individual sugars Gluc g	Fruct g	Sucr g	Malt g	Lact g	Dietary fibre Southgate method g	Englyst method g	Fatty acids Satd g	Mono-unsatd g	Poly-unsatd g	Cholesterol mg
	Sugars and syrups													
49	**Glucose liquid**, BP[a]	N	40.2	N	0	0	N	0	0	0	0	0	0	0
50	**Honey**	0	76.4	34.6	41.8	Tr	Tr	0	0	0	0	0	0	0
51	**Honeycomb**	0	74.4	34.2	40.2	0	0	0	0	0	0	0	0	0
52	**Ice Magic sauce**	2.1	29.4	Tr	Tr	29.4	0	0	0	0	N	N	N	Tr
53	**Ice cream sauce**, topping	2.9	51.0	17.1	15.9	18.0	0	0	0	0	Tr	Tr	Tr	0
54	**Icing**, butter	0.7	63.3	0.1	0.1	63.0	0	Tr	0	0	19.6	7.0	1.0	82
55	fondant	0.8	89.8	0.8	Tr	89.0	Tr	0	0	0	0	0	0	0
56	glacé	0.9	89.0	0	0	89.0	0	0	0	0	0	0	0	0
57	Royal	1.0	94.1	Tr	Tr	94.1	0	0	0	0	0	0	0	0
58	**Jaggery**	7.9	89.3	N	N	N	N	0	0	0	0	0	0	0
59	**Molasses**	0	59.4	11.5	12.9	35.1	Tr	0	Tr	Tr	Tr	Tr	Tr	0
60	**Sugar**, brown	0	101.3[b]	0	0	101.3	0	0	0	0	0	0	0	0
61	Demerara	0	104.5	0	0	104.5	0	0	0	0	0	0	0	0
62	icing	1.1	103.8	0	0	103.8	0	0	0	0	0	0	0	0
63	white	0	105.0	0	0	105.0	0	0	0	0	0	0	0	0
64	**Syrup**, corn dark	N	28.1	14.9	1.2	2.2	9.8	0	0	0	0	0	0	0
65	golden	0	79.0	23.1	23.0	32.8	0	0	0	0	0	0	0	0
66	-, pouring	0	79.0	23.1	23.0	32.8	0	0	0	0	0	0	0	0
67	maple	0	(60.7)	3.1	1.1	56.5	Tr	0	0	0	Tr	Tr	Tr	0
68	**Treacle**, black	0	67.2	17.4	16.7	32.7	0	0	Tr	Tr	0	0	0	0

[a] Proportion of glucose varies with the dextrose equivalent
[b] Light muscovado sugar contains 100.3g per 100g. Dark muscovado sugar contains 94.5g per 100g

Inorganic constituents per 100g food

No. 17-	Food	Na (mg)	K (mg)	Ca (mg)	Mg (mg)	P (mg)	Fe (mg)	Cu (mg)	Zn (mg)	Cl (mg)	Mn (mg)	Se (µg)	I (µg)
	Sugars and syrups												
49	**Glucose liquid**, BP	150	3	8	2	11	0.5	0.09	N	190	Tr	Tr	Tr
50	**Honey**	11	51	5	2	17	0.4	0.05	0.9	18	0.30	(1)	Tr
51	**Honeycomb**	7	35	8	2	32	0.2	0.04	N	26	N	(1)	Tr
52	**Ice Magic sauce**	40	270	140	45	130	4.6	0.23	0.8	100	0.27	N	N
53	**Ice cream sauce**, topping	140	68	9	15	26	0.8	0.09	0.1	40	0.14	N	N
54	**Icing**, butter	280	14	7	1	1	0.4	0.04	0.1	410	Tr	Tr	14
55	fondant	29	24	(9)	3	5	(0.2)	0.11	(0.1)	23	Tr	1	Tr
56	glacé	10	9	2	Tr	Tr	0.4	0.04	Tr	Tr	Tr	Tr	Tr
57	Royal	29	25	2	1	4	0.5	0.05	0.1	16	Tr	1	Tr
58	**Jaggery**	79	290	92	120	72	1.6	0.75	0.1	250	0.50	Tr	Tr
59	**Molasses**	37	1460	210	240	31	4.7	0.49	0.3	N	1.53	Tr	Tr
60	**Sugar**, brown	31	140	56	17	4	1.7	0.06	0.1	N	Tr	Tr	Tr
61	Demerara	5	48	29	9	3	0.9	0.11	(0.1)	35	Tr	Tr	Tr
62	icing	12	11	2	Tr	1	0.5	0.05	0.1	Tr	Tr	Tr	Tr
63	white	5	5	(10)	(2)	(1)	(0.2)	0.12	(0.1)	Tr	Tr	Tr	Tr
64	**Syrup**, corn, dark	160	44	18	8	11	0.4	Tr	Tr	N	Tr	Tr	Tr
65	golden	270	58	16	3	(1)	0.4	0.06	(0.1)	42	0.01	Tr	Tr
66	-, pouring	340	71	13	1	Tr	0.5	0.04	Tr	N	0.01	Tr	Tr
67	maple	9	200	67	14	2	1.2	0.07	4.2	N	3.30	Tr	Tr
68	**Treacle**, black	180	1760	550	180	29	21.3	0.78	0.8	820	2.67	N	Tr

No. 17-	Food	Retinol µg	Carotene µg	Vitamin D µg	Vitamin E mg	Thiamin mg	Riboflavin mg	Niacin mg	Trypt/60 mg	Vitamin B_6 mg	Vitamin B_{12} µg	Folate µg	Pantothenate mg	Biotin µg	Vitamin C mg
	Sugars and syrups														
49	**Glucose liquid**, BP	0	0	0	0	0	0	0	0	0	0	0	0	0	0
50	**Honey**	0	0	0	0	Tr	0.05	0.2	Tr	N	0	N	N	N	0
51	**Honeycomb**	0	0	0	0	Tr	0.05	0.2	Tr	N	0	N	N	N	0
52	**Ice Magic sauce**	Tr	Tr	0	N	Tr	Tr	Tr	Tr	Tr	Tr	Tr	Tr	Tr	0
53	**Ice cream sauce**, topping	0	Tr	0	N	Tr	Tr	Tr	Tr	Tr	0	Tr	Tr	Tr	0
54	**Icing**, butter	290	155	0.3	0.71	Tr	0.01	Tr	Tr	Tr	Tr	Tr	0.02	Tr	Tr
55	fondant	0	Tr	0	Tr	Tr	0.05	Tr	0.3	Tr	Tr	2	0.04	1	1
56	glacé	0	0	0	0	0	0	0	0	0	0	0	0	0	0
57	Royal	0	Tr	0	Tr	Tr	0.04	Tr	0.3	Tr	Tr	1	0.03	1	Tr
58	**Jaggery**	0	0	0	N	Tr	0.04	Tr	Tr	Tr	0	Tr	Tr	Tr	0
59	**Molasses**	0	0	0	0.41	0.04	Tr	0.9	0	0.67	0	0	0.80	Tr	0
60	**Sugar**, brown	0	0	0	0	Tr	Tr	Tr	Tr	Tr	0	Tr	Tr	Tr	0
61	Demerara	0	0	0	0	Tr	Tr	Tr	Tr	Tr	0	Tr	Tr	Tr	0
62	icing	0	0	0	0	0	0	0	0	0	0	0	0	0	0
63	white	0	0	0	0	0	0	0	0	0	0	0	0	0	0
64	**Syrup**, corn, dark	0	0	0	0	0.01	0.01	0	0	0.01	0	0	0.02	0	0
65	golden	0	0	0	0	Tr	Tr	Tr	Tr	Tr	0	Tr	Tr	Tr	0
66	-, pouring	0	0	0	0	Tr	Tr	Tr	Tr	Tr	0	Tr	Tr	Tr	0
67	maple	0	0	0	0	0.01	0.01	Tr	Tr	Tr	0	0	0.04	Tr	0
68	**Treacle**, black	0	0	0	0	Tr	Tr	Tr	Tr	Tr	0	Tr	Tr	Tr	0

No. 17-	Food	Description and main data sources	Water g	Total Nitrogen g	Protein g	Fat g	Carbo-hydrate g	Energy value kcal	Energy value kJ
	Preserves								
69	**Chocolate spread**	6 samples, 3 brands	0.2	0.66	4.1	37.6	57.1	569	2375
70	**Chocolate nut spread**	8 samples, 5 brands	Tr	0.99	6.2	33.0	60.5	549	2294
71	**Fruit spread**	8 samples, 4 brands; assorted flavours	64.0	0.11	0.7	0.1	31.4	121	518
72	**Jam**, diabetic	Analysis and manufacturers' data	N	0.02	0.1	0	60.4[a]	174	728
73	fruit with edible seeds	10 samples, 5 flavours	29.8	0.10	0.6	0	69.0	261	1114
74	stone fruit	8 samples, 4 flavours	29.6	0.06	0.4	0	69.3	261	1116
75	reduced sugar	9 samples, 5 brands; assorted flavours	65.3	0.08	0.5	0.1	31.9	123	523
76	**Lemon curd**	10 jars, 4 brands	30.7	0.09	0.6	5.0	62.7	283	1198
77	homemade	Recipe	15.4	0.67	4.2	16.7	51.2	358	1506
78	**Marmalade**	4 brands	28.0	0.01	0.1	0	69.5[b]	261	1114
79	diabetic	Manufacturers' data	N	0.02	0.1	0	57.0[c]	173	725
80	**Mincemeat**	10 samples of the same brand	27.5	0.10	0.6	4.3	62.1	274	1163
81	vegetarian	Recipe	34.5	0.20	1.2	12.6	43.7	305	1281
	Chocolate confectionery								
82	**Bounty bar**	8 samples; plain and milk chocolate	7.6	0.77	4.8	26.1	58.3	473	1980
83	**Chocolate covered caramels**	18 samples, 4 brands including Rolo, Caramel	5.6	0.80	5.0	21.7	66.5[d]	465	1952
84	**Chocolate covered bar with fruit/ nuts and wafer/biscuit**	28 samples of Lion Bar, Picnic, Ballisto, Crispy Caramel	2.7	1.39	8.7	27.4	58.0[e]	499	2090

[a] Contains 7g carbohydrate and 53.4g sorbitol per 100g
[b] Reduced sugar marmalade contains about 33g carbohydrate
[c] Contains 3g carbohydrate and 54g sorbitol per 100g
[d] Includes 10.7g maltodextrins
[e] Includes 2.6g maltodextrins

Carbohydrate fractions and fatty acids, g per 100g food
Cholesterol, mg per 100g food

No.	Food	Starch	Total sugars	Individual sugars					Dietary fibre		Fatty acids			Cholesterol
17-				Gluc	Fruct	Sucr	Malt	Lact	Southgate method	Englyst method	Satd	Mono-unsatd	Poly-unsatd	
		g	g	g	g	g	g	g	g	g	g	g	g	mg
	Preserves													
69	**Chocolate spread**	Tr	57.1	Tr	Tr	53.1	Tr	4.0	N	N	N	N	N	(2)
70	**Chocolate nut spread**	0.8	59.7	Tr	Tr	56.7	0	3.0	1.2	0.8	10.1	16.8	4.6	2
71	**Fruit spread**	0.7	30.7	13.6	16.7	0.4	0	0	N	N	Tr	Tr	Tr	0
72	**Jam**, diabetic	0	7.0	2.9	3.9	0.2	0	0	(1.0)	N	0	0	0	0
73	fruit with edible seeds	0	69.0	27.4	14.9	18.7	8.0	0	1.0	N	0	0	0	0
74	stone fruit	0	69.3	27.5	14.9	18.8	8.0	0	0.9	N	0	0	0	0
75	reduced sugar	0	31.9	10.4	15.0	6.5	0	0	N	(0.8)	Tr	Tr	Tr	0
76	**Lemon curd**, starch base	22.3	40.4	16.5	7.6	12.0	4.3	0	0.2	(0.2)	1.6	2.1	1.2	21
77	homemade	0	51.2	0.1	0.2	50.8	0	Tr	Tr	Tr	9.8	4.7	0.8	160
78	**Marmalade**	0	69.5	27.6	15.0	18.8	8.0	0	0.6	(0.3)	0	0	0	0
79	diabetic	0	N	N	N	N	0	0	(0.6)	(0.6)	0	0	0	0
80	**Mincemeat**	Tr	62.1	30.7	30.8	0.6	Tr	0	3.0	1.3	N	N	N	4
81	vegetarian	1.3	42.4	12.4	10.8	15.7	3.4	0	(2.1)	(1.4)	N	N	N	0
	Chocolate confectionery													
82	**Bounty bar**	4.6	53.7	3.5	0.1	44.2	3.2	2.6	N	N	21.2	3.2	0.4	10
83	**Chocolate covered caramels**	Tr	55.8	5.2	2.9	36.8	4.3	7.0	N	N	10.7	9.1	0.7	23
84	**Chocolate covered bar with fruit/nuts and wafer/biscuit**	9.9	45.5	4.4	2.2	31.3	2.3	5.3	N	3.8	13.4	10.5	2.2	11

Inorganic constituents per 100g food

No. 17-	Food	mg										µg	
		Na	K	Ca	Mg	P	Fe	Cu	Zn	Cl	Mn	Se	I
	Preserves												
69	**Chocolate spread**	N	N	N	N	N	N	N	N	N	N	N	N
70	**Chocolate nut spread**	50	390	130	65	180	2.2	0.48	1.0	60	1.10	N	N
71	**Fruit spread**	10	190	11	10	20	0.6	Tr	0.1	10	0.20	Tr	N
72	**Jam**, diabetic	(38)	(55)	(11)	(4)	(8)	(0.2)	(0.02)	Tr	(7)	(0.08)	Tr	(7)
73	fruit with edible seeds	29	43	12	5	10	0.2	0.01	(0.1)	9	0.13	Tr	7
74	stone fruit	46	67	10	3	6	0.2	0.02	Tr	4	0.02	Tr	7
75	reduced sugar	20	120	19	7	15	0.4	0.05	Tr	Tr	0.10	0	(2)
76	**Lemon curd**, starch base	65	11	9	2	15	0.5	(0.30)	1.3	150	N	N	N
77	homemade	170	78	27	7	71	0.8	0.10	0.5	240	Tr	4	(23)
78	**Marmalade**	64	35	26	3	6	0.2	0.03	(0.1)	7	0.01	(1)	(7)
79	diabetic	(64)	(35)	(26)	(3)	(6)	(0.2)	(0.03)	(0.1)	(7)	(0.01)	(1)	(7)
80	**Mincemeat**	18	44	35	4	13	0.6	0.12	0.2	7	N	(1)	(7)
81	vegetarian	50	300	71	18	31	1.3	0.23	0.2	10	0.35	Tr	Tr
	Chocolate confectionery												
82	**Bounty bar**	180	320	110	43	140	1.3	0.47	N	400	N	N	N
83	**Chocolate covered caramels**	180	270	420	31	150	1.1	0.02	0.6	260	0.12	N	N
84	**Chocolate covered bar with fruit/nuts and wafer/biscuit**	160	330	130	50	190	1.7	0.24	1.0	250	0.59	N	N

No. 17-	Food	Retinol µg	Carotene µg	Vitamin D µg	Vitamin E mg	Thiamin mg	Riboflavin mg	Niacin mg	Trypt/60 mg	Vitamin B_6 mg	Vitamin B_{12} µg	Folate µg	Pantothenate mg	Biotin µg	Vitamin C mg
	Preserves														
69	**Chocolate spread**	Tr	Tr	Tr	N	N	N	N	N	N	Tr	N	N	N	0
70	**Chocolate nut spread**	Tr	Tr	Tr	N	0.03	0.10	0.5	1.5	0.10	Tr	N	N	N	Tr
71	**Fruit spread**	0	Tr	0	Tr	Tr	Tr	Tr	Tr	Tr	0	Tr	Tr	Tr	6
72	**Jam**, diabetic	0	Tr	0	0	Tr	Tr	Tr	Tr	Tr	0	Tr	Tr	Tr	Tr
73	fruit with edible seeds	0	Tr	0	0	Tr	Tr	Tr	Tr	Tr	0	Tr	Tr	Tr	10[a]
74	stone fruit	0	Tr	0	0	Tr	Tr	Tr	Tr	Tr	0	Tr	Tr	Tr	0
75	reduced sugar	0	(26)	0	(0.14)	Tr	Tr	Tr	Tr	Tr	0	Tr	Tr	Tr	26
76	**Lemon curd**, starch base	(10)	Tr	(0.1)	N	Tr	(0.02)	Tr	(0.1)	Tr	Tr	Tr	(0.10)	(1)	Tr
77	homemade	195	72	0.7	0.68	0.03	0.16	0	1.2	0.04	1	10	0.48	7	4
78	**Marmalade**	0	50	0	Tr	Tr	Tr	Tr	Tr	Tr	0	5	Tr	Tr	10
79	diabetic	0	(50)	0	Tr	Tr	Tr	Tr	Tr	Tr	0	(5)	Tr	Tr	(10)
80	**Mincemeat**	0	9	Tr	N	0.04	0.02	0.4	0.1	(0.10)	Tr	8	0.03	Tr	Tr
81	vegetarian	0	(12)	0	N	0.05	0.02	0.3	0.1	0.08	0	5	0.04	2	4
	Chocolate confectionery														
82	**Bounty bar**	N	(40)	Tr	N	(0.04)	(0.10)	(0.3)	0.8	(0.03)	Tr	N	(0.59)	(2)	0
83	**Chocolate covered caramels**	33	20	Tr	2.37	0.06	0.34	0.3	1.1	0.02	Tr	4	0.60	3	0
84	**Chocolate covered bar with fruit/nuts and wafer/biscuit**	20	18	0	2.91	0.11	0.26	2.3	2.4	0.11	0	17	N	N	0

[a] Blackcurrant jam contains 24mg vitamin C per 100g

No. 17-	Food	Description and main data sources	Water g	Total Nitrogen g	Protein g	Fat g	Carbo-hydrate g	Energy value kcal	Energy value kJ
	***Chocolate confectionery** continued*								
85	**Chocolate covered ice cream bar**	10 samples including Mars, Bounty, Snickers, Milky Way	33.1	0.78	5.0	23.3	24.0	320	1331
86	**Chocolate**, cooking	10 samples	0.3	0.70	4.8	34.8	57.8	549	2294
87	diabetic	10 samples, 2 brands	2.2	1.49	9.3	29.5	38.6	447	1867
88	fancy and filled	10 samples of different brands	6.1	0.78	4.9	21.3	62.9[a]	447	1878
89	milk	12 bars, 5 brands including Dairy Milk, Galaxy, chocolate buttons	1.3	1.23	7.7	30.7	56.9	520	2177
90	plain	6 bars, 3 brands	0.6	0.80	5.0	28.0	63.5	510	2137
91	white	14 samples, 5 brands; buttons and bars	0.6	1.28	8.0	30.9	58.3	529	2212
92	**Creme eggs**	10 samples	6.7	0.66	4.1	14.6	*71.0*[b]	*417*	*1746*
93	**Kit Kat**	Analysis and manufacturer's data	2.0	1.20	7.5	26.0	63.0	500	2098
94	**Mars bar**	8 samples	6.9	0.84	5.3	18.9	66.5	441	1853
95	**Milky Way**	10 samples	6.6	0.70	4.4	15.8	63.4	398	1674
96	**Smartie-type sweets**	10 samples including Smarties and M & M's	1.5	0.86	5.4	17.5	73.9	456	1922
97	**Snickers**	Ref. Cutrufelli and Pehrsson (1991)	5.9	1.68	9.6	22.3	*60.3*	*455*	*1904*
98	**Truffles**, mocha	Recipe	2.6	0.99	6.2	24.9	63.8	488	2049
99	rum	Recipe	3.4	0.97	6.1	33.7	49.7	521[c]	2175[c]
100	**Twix**	10 samples	3.5	0.90	5.6	24.5	63.2	480	2013

[a] Includes 2.7g maltodextrins
[b] Includes 16g maltodextrins
[c] Includes contribution from 1.1g alcohol

No.	Food			Individual sugars					Dietary fibre		Fatty acids			
17-		Starch	Total sugars	Gluc	Fruct	Sucr	Malt	Lact	Southgate method	Englyst method	Satd	Mono-unsatd	Poly-unsatd	Cholest-erol
		g	g	g	g	g	g	g	g	g	g	g	g	mg
	***Chocolate confectionery** continued*													
85	**Chocolate covered ice cream bar**	0.6	23.4	2.2	0.1	18.9	Tr	2.2	N	N	(15.1)	(6.7)	(1.5)	(9)
86	**Chocolate**, cooking	1.9	55.9	Tr	Tr	55.9	Tr	Tr	N	N	28.0	4.6	0.7	0
87	diabetic	N[a]	19.1	Tr	N[a]	Tr	Tr	10.3	N	N	(18.2)	(9.7)	(1.5)	29
88	fancy and filled	0.2	60.0	5.4	3.1	45.7	2.2	3.6	N	1.3	11.3	8.0	1.0	11
89	milk	Tr	56.9	0.1	0.1	46.6	Tr	10.1	N	0.8	18.3	9.9	1.2	23
90	plain	0.9	62.6	Tr	Tr	62.4	Tr	0.2	N	2.5	16.8	9.0	1.0	6
91	white	Tr	58.3	Tr	Tr	47.6	Tr	10.7	N	N	18.4	10.0	1.1	N
92	**Creme eggs**	Tr	58.0	3.6	1.8	45.7	2.0	4.9	N	N	4.7	4.7	0.5	10
93	**Kit Kat**	12.9	50.1	0.1	0.1	42.2	0.1	7.6	N	1.2	17.7	7.1	0.9	12
94	**Mars bar**	0.7	65.8	10.0	0.1	42.5	6.7	6.5	N	N	10.0	7.2	0.8	25
95	**Milky Way**	0.8	62.6	8.3	Tr	43.3	4.9	6.1	N	N	8.3	5.8	0.9	N
96	**Smartie-type sweets**	3.1	70.8	0.3	Tr	65.6	0.1	4.8	N	N	10.4	5.7	0.6	(17)
97	**Snickers**	N	44.2	6.4	0.5	26.8	6.2	4.3	N	N	12.0	6.7	0.9	N
98	**Truffles**, mocha	0.8	62.9	Tr	Tr	60.1	Tr	2.8	N	2.0	15.0	7.9	0.9	13
99	rum	0.7	49.0	Tr	Tr	48.6	Tr	0.4	N	1.9	19.5	10.6	1.5	165
100	**Twix**	15.5	47.7	6.6	0.1	31.2	3.9	5.9	N	N	12.2	9.6	1.7	N

[a] Some samples contain fructose and some contain isomalt

Inorganic constituents per 100g food

No. 17-	Food	mg										µg	
		Na	K	Ca	Mg	P	Fe	Cu	Zn	Cl	Mn	Se	I
	***Chocolate confectionery** continued*												
85	**Chocolate covered ice cream bar**	91	250	140	31	150	0.7	0.04	0.7	180	0.16	N	16
86	**Chocolate**, cooking	130	N	N	N	N	N	N	N	N	N	N	N
87	diabetic	(65)	(350)	(180)	(55)	(200)	(1.3)	(0.30)	(1.1)	(150)	(0.30)	(4)	N
88	fancy and filled	88	270	110	48	150	1.2	0.30	0.8	140	0.39	(2)	120
89	milk	85	390	220	50	220	1.4	0.24	1.1	190	0.22	(4)	30
90	plain	6	300	33	89	140	2.3	0.71	1.3	9	0.63	(4)	3
91	white	110	350	270	26	230	0.2	Tr	0.9	250	0.02	N	N
92	**Creme eggs**	55	210	120	27	130	0.8	0.10	0.6	110	0.10	Tr	N
93	**Kit Kat**	120	330	200	52	200	1.5	0.28	1.1	210	0.34	N	N
94	**Mars bar**	150	250	160	35	150	1.1	0.31	N	300	N	(2)	N
95	**Milky Way**	100	240	120	42	140	1.7	0.13	0.8	160	0.25	N	N
96	**Smartie-type sweets**	58	280	150	48	160	1.5	0.25	0.9	120	0.25	N	N
97	**Snickers**	270	330	120	60	210	0.8	0.24	1.2	N	0.49	N	N
98	**Truffles**, mocha	37	390	92	85	170	2.0	0.59	1.3	58	0.55	(2)	18
99	rum	78	260	48	72	180	2.6	0.57	1.5	130	0.50	(4)	(23)
100	**Twix**	190	190	110	28	130	1.1	0.08	0.7	250	0.22	N	N

No. 17-	Food	Retinol µg	Carotene µg	Vitamin D µg	Vitamin E mg	Thiamin mg	Riboflavin mg	Niacin mg	Trypt/60 mg	Vitamin B_6 mg	Vitamin B_{12} µg	Folate µg	Pantothenate mg	Biotin µg	Vitamin C mg
	Chocolate confectionery *continued*														
85	**Chocolate covered ice cream bar**	78	47	0.2	N	0.05	0.29	0.6	1.4	0.04	0	12	0.50	7	0
86	**Chocolate**, cooking	N	N	Tr	N	N	N	N	N	N	Tr	N	N	N	0
87	diabetic	N	(39)	Tr	(0.72)	(0.10)	(0.23)	(0.2)	(1.4)	(0.07)	Tr	(10)	(1.08)	(3)	0
88	fancy and filled	81	120	Tr	1.65	0.05	0.20	0.4	0.9	0.03	Tr	17	(0.73)	(3)	0
89	milk	25	11	0	0.45	0.07	0.49	0.4	2.3	0.04	1	11	0.70	4	0
90	plain	15	15	0	1.44	0.04	0.06	0.4	0.7	0.03	0	12	0.30	3	0
91	white	13	75	Tr	1.14	0.08	0.49	0.2	2.6	0.07	Tr	(10)	(0.59)	3	0
92	**Creme eggs**	47	(55)	0.6	1.07	0.06	0.34	0.2	1.3	0.03	1	12	(0.59)	(3)	0
93	**Kit Kat**	8	47	Tr	1.03	0.11	0.44	0.5	2.6	0.06	Tr	N	0.70	4	0
94	**Mars bar**	N	(40)	Tr	N	(0.05)	(0.20)	(0.2)	0.9	(0.03)	Tr	N	(0.59)	(2)	0
95	**Milky Way**	N	Tr	Tr	1.91	0.05	0.20	0.2	1.1	0.03	Tr	(10)	0.59	2	0
96	**Smartie-type sweets**	5	28	Tr	0.80	0.08	0.79	0.3	1.7	0.03	Tr	4	0.67	2	0
97	**Snickers**	N	N	Tr	N	0.05	0.18	3.0	N	0.19	0	40	0.58	N	0
98	**Truffles**, mocha	36	27	1.2	1.21	0.05	0.15	0.8	1.1	0.04	0	13	0.43	4	1
99	rum	165	55	0.7	1.70	0.07	0.12	0.3	1.2	0.06	1	26	0.82	9	Tr
100	**Twix**	1	7	Tr	3.72	0.06	0.22	0.6	0.9	0.05	Tr	N	0.61	3	0

Composition of food per 100g

No. 17-	Food	Description and main data sources	Water g	Total Nitrogen g	Protein g	Fat g	Carbo-hydrate g	Energy value kcal	Energy value kJ
	Non chocolate confectionery								
101	**Boiled sweets**	6 samples	(16.6)	Tr	Tr	Tr	87.1	327	1394
102	**Cereal chewy bar**	17 bars of different brands; assorted types	1.1	1.17	7.3	16.4	64.7[a]	419	1766
103	**Cereal crunchy bar**	12 bars of different brands; assorted types	2.6	1.66	10.4	22.2	60.5[b]	468	1966
104	**Chew sweets**	15 samples, 6 brands including Opal Fruits, Chewitts, Fruit-tella	7.6	0.16	1.0	5.6	87.0[c]	381	1616
105	**Coconut ice**	Recipe	17.3	0.27	1.7	12.7	66.7	371	1566
106	**Foam sweets**	7 samples	6.5	0.70	4.4	4.8	80.1[d]	361	1534
107	**Fruit gums/jellies**	11 samples, 10 brands; assorted flavours	14.0	1.04	6.5	0	79.5[e]	324	1383
108	**Fruit pastilles**	6 samples of different brands; assorted flavours	9.1	0.45	2.8	0	84.2[f]	327	1395
109	**Fudge**	Recipe	4.0	0.53	3.3	13.7	81.1	441	1860
110	**Halva**, carrot	Recipe	30.6	0.73	4.5	18.9	44.2	354	1483
111	semolina	Recipe	35.0	0.35	2.2	16.6	49.0	342	1435
112	**Liquorice allsorts**	7 samples, 4 brands	8.4	0.59	3.7	5.2	76.7[g]	349	1483
113	**Liquorice shapes**	9 samples, 4 brands	13.3	0.88	5.5	1.4	65.0[h]	278	1185
114	**Marshmallows**	7 samples of different brands	17.4	0.62	3.9	0	83.1[i]	327	1396
115	**Nougat**	Recipe	(13.3)	0.70	4.4	8.5	77.3	384	1626
116	**Peanut brittle**	Recipe	(1.7)	1.38	8.6	19.0	73.8	483	2031
117	**Peppermints**	Several samples of 6 different brands	0.2	0.08	0.5	0.7	102.7	393	1678
118	**Peppermint creams**	Recipe	6.7	0.09	0.6	Tr	97.9	369	1576
119	**Sherbert sweets**	10 samples of different brands	0.2	0.10	0.6	0	93.9	355	1513

[a] Includes 6.4g maltodextrins
[b] Includes 4.4g maltodextrins
[c] Includes 32.0g maltodextrins
[d] Includes 7.8g maltodextrins
[e] Includes 18.9g maltodextrins
[f] Includes 21.5g maltodextrins
[g] Includes 4.9g maltodextrins
[h] Includes 3.3g maltodextrins
[i] Includes 14.1g maltodextrins

Carbohydrate fractions and fatty acids, g per 100g food
Cholesterol, mg per 100g food

No. 17-	Food	Starch	Total sugars	Individual sugars					Dietary fibre		Fatty acids			Cholesterol
				Gluc	Fruct	Sucr	Malt	Lact	Southgate method	Englyst method	Satd	Mono-unsatd	Poly-unsatd	
		g	g	g	g	g	g	g	g	g	g	g	g	mg
	Non chocolate confectionery													
101	**Boiled sweets**	0.4	86.7	8.5	1.4	67.5	9.3	0	0	0	0	0	0	0
102	**Cereal chewy bar**	25.6	32.7	7.4	5.0	10.6	8.2	1.5	N	3.2	N	N	N	N
103	**Cereal crunchy bar**	28.2	27.9	2.3	1.5	21.8	1.5	0.8	N	4.8	4.5	11.3	5.4	Tr
104	**Chew sweets**	Tr	55.0	7.7	0.8	39.4	7.1	Tr	N	1.0	3.0	2.2	0.2	0
105	**Coconut ice**	Tr	66.7	Tr	0.2	65.7	Tr	0.9	4.1	2.6	10.7	0.9	0.3	3
106	**Foam sweets**	Tr	72.3	4.7	2.0	60.0	5.6	Tr	N	0.9	N	N	N	0
107	**Fruit gums/jellies**	1.9	58.7	6.3	Tr	46.4	6.0	Tr	N	N	0	0	0	0
108	**Fruit pastilles**	3.4	59.3	6.5	2.1	45.4	5.3	Tr	N	N	0	0	0	0
109	**Fudge**	Tr	81.1	Tr	Tr	77.4	Tr	3.6	0	0	9.0	3.5	0.4	41
110	**Halva**, carrot	0.4	43.8	2.6	2.2	34.1	0	4.9	(3.0)	3.0	11.2	5.1	1.3	50
111	semolina	11.4	37.6	Tr	Tr	37.6	0	0	(0.7)	(0.5)	9.8	4.5	1.0	41
112	**Liquorice allsorts**	9.4	62.4	5.9	2.5	51.1	2.9	Tr	N	2.0	3.6	1.2	0.2	0
113	**Liquorice shapes**	20.2	41.5	5.0	5.4	27.7	3.4	Tr	N	1.9	N	N	N	0
114	**Marshmallows**	4.5	64.5	12.1	0.7	41.5	10.2	Tr	0	0	0	0	0	0
115	**Nougat**	0.4	76.9	Tr	3.8	73.1	Tr	0	N	0.9	1.1	4.2	2.7	0
116	**Peanut brittle**	2.1	71.8	3.8	3.8	64.2	0	0	2.4	2.0	5.3	7.9	4.8	11
117	**Peppermints**	0	102.7	1.0	0	101.7	0	0	0	0	N	N	N	0
118	**Peppermint creams**	1.0	96.8	Tr	Tr	96.8	0	0	Tr	Tr	0	0	0	0
119	**Sherbert sweets**	Tr	93.9	0.2	Tr	93.7	Tr	Tr	Tr	Tr	0	0	0	0

Inorganic constituents per 100g food

No.	Food	mg										µg	
17-		Na	K	Ca	Mg	P	Fe	Cu	Zn	Cl	Mn	Se	I
	Non chocolate confectionery												
101	**Boiled sweets**	25	8	5	2	12	0.4	0.09	N	68	N	Tr	N
102	**Cereal chewy bar**	110	320	70	55	190	1.9	0.16	1.1	210	1.36	N	N
103	**Cereal crunchy bar**	74	360	77	86	290	2.6	0.29	1.7	140	2.09	N	N
104	**Chew sweets**	48	15	6	4	4	0.2	Tr	Tr	67	Tr	N	N
105	**Coconut ice**	16	160	28	19	48	0.7	0.12	0.4	58	0.35	1	3
106	**Foam sweets**	35	38	41	9	29	0.4	0.04	0.1	55	0.16	N	N
107	**Fruit gums/jellies**	30	8	5	1	4	0.1	0.02	Tr	N	Tr	Tr	N
108	**Fruit pastilles**	33	28	28	6	4	0.4	0.04	Tr	29	0.02	Tr	N
109	**Fudge**	160	140	120	13	100	(0.3)	0.10	0.4	240	Tr	(1)	12
110	**Halva**, carrot	100	370	150	19	130	0.6	0.08	0.6	170	0.14	2	(18)
111	semolina	13	56	12	12	32	0.3	0.09	(0.2)	28	(0.13)	Tr	N
112	**Liquorice allsorts**	57	600	170	76	44	7.3	0.34	0.5	N	1.14	N	N
113	**Liquorice shapes**	190	1590	440	170	74	16.7	0.72	1.0	990	2.49	N	N
114	**Marshmallows**	29	2	4	2	4	0.3	Tr	Tr	36	Tr	N	N
115	**Nougat**	120	240	25	23	73	0.7	0.22	0.5	160	0.17	(2)	N
116	**Peanut brittle**	110	240	32	72	140	1.1	0.41	1.2	69	0.69	1	8
117	**Peppermints**	9	Tr	7	3	Tr	0.2	0.04	N	22	N	Tr	N
118	**Peppermint creams**	24	21	2	1	3	0.5	0.05	0.1	11	Tr	Tr	Tr
119	**Sherbert sweets**	1050	15	42	69	Tr	0.2	0.04	Tr	6	0.02	N	N

No. 17-	Food	Retinol µg	Carotene µg	Vitamin D µg	Vitamin E mg	Thiamin mg	Riboflavin mg	Niacin mg	Trypt/60 mg	Vitamin B_6 mg	Vitamin B_{12} µg	Folate µg	Pantothenate mg	Biotin µg	Vitamin C mg
	Non chocolate confectionery														
101	**Boiled sweets**	0	0	0	0	0	0	0	0	0	0	0	0	0	0
102	**Cereal chewy bar**	0	N	0	N	0.24	0.17	1.2	1.9	0.13	Tr	11	N	N	Tr
103	**Cereal crunchy bar**	0	Tr	0	3.84	0.24	0.12	2.3	4.4	0.14	0	15	N	N	Tr
104	**Chew sweets**	0	315	0	0.91	Tr	Tr	N	N	Tr	0	Tr	Tr	Tr	0
105	**Coconut ice**	10	4	0	0.21	0.01	0.04	0.2	0.4	0.03	Tr	2	0.15	N	Tr
106	**Foam sweets**	Tr	Tr	0	0.02	0	0	0.1	0	0.01	0	1	0	0	0
107	**Fruit gums, jellies**	0	N	0	0	0	0	0	0	0	0	0	0	0	0
108	**Fruit pastilles**	0	0	0	0	0	0	0	0	0	0	0	0	0	0
109	**Fudge**	145	87	1.2	0.26	0.02	0.15	0.1	0.8	0.03	0	4	0.28	2	0
110	**Halva**, carrot	140	9430	0.3	1.29	0.11	0.15	0.3	1.0	(0.13)	0	13	0.65	3	Tr
111	semolina	98	75	0.3	0.94	(0.02)	0.01	0.1	0.4	(0.01)	Tr	(2)	(0.03)	1	0
112	**Liquorice allsorts**	0	0	0	0	0	0	0	0.2	0	0	0	0	0	0
113	**Liquorice shapes**	0	0	0	0	0.10	0.08	1.5	0.6	0.56	0	9	N	N	0
114	**Marshmallows**	0	0	0	0	0	0	Tr	Tr	0	0	0	0	0	0
115	**Nougat**	0	20	0	0.64	0.11	0.12	0.3	1.1	Tr	0	11	(0.06)	N	0
116	**Peanut brittle**	38	20	0	3.42	0.38	0.03	4.5	1.8	0.19	0	36	0.88	24	0
117	**Peppermints**	0	0	0	0	0	0	0	0	0	0	0	0	0	0
118	**Peppermint creams**	0	Tr	0	Tr	Tr	0.03	Tr	0.2	Tr	0	1	0.02	Tr	0
119	**Sherbert sweets**	0	0	0	0	0	0	Tr	Tr	0	0	0	0	0	0

No. 17-	Food	Description and main data sources	Water g	Total Nitrogen g	Protein g	Fat g	Carbo-hydrate g	Energy value kcal	Energy value kJ
	Non chocolate confectionery *continued*								
120	**Toffees**, mixed	13 samples, 4 brands including cream and plain varieties	2.4	0.35	2.2	18.6	66.7[a]	426	1793
121	**Turkish delight**, with nuts	Recipe	13.6	0.66	4.1	2.6	81.1	344	1462
122	without nuts	7 assorted samples	16.1	0.10	0.6	0	77.9	295	1257

[a] Includes 21.9g maltodextrins

Carbohydrate fractions and fatty acids, g per 100g food
Cholesterol, mg per 100g food

No. 17-	Food	Starch g	Total sugars g	Individual sugars: Gluc g	Fruct g	Sucr g	Malt g	Lact g	Dietary fibre: Southgate method g	Englyst method g	Fatty acids: Satd g	Mono-unsatd g	Poly-unsatd g	Cholesterol mg
	Non chocolate confectionery *continued*													
120	**Toffees**, mixed	Tr	44.8	5.2	0.7	32.4	4.5	2.0	0	0	9.5	7.5	0.7	17
121	**Turkish delight**, with nuts	0.2	80.9	Tr	Tr	80.9	0	0	N	0.3	0.3	1.3	0.8	0
122	without nuts	9.3	68.6	N	N	N	N	0	0	0	0	0	0	0

Inorganic constituents per 100g food

No.	Food	mg										µg	
17-		Na	K	Ca	Mg	P	Fe	Cu	Zn	Cl	Mn	Se	I
	Non chocolate confectionery *continued*												
120	**Toffees**, mixed	340	110	73	8	62	0.2	0.02	0.3	500	Tr	N	N
121	**Turkish delight**, with nuts	84	53	22	8	21	0.4	0.13	0.2	(37)	0.05	(1)	Tr
122	without nuts	31	4	10	2	7	0.2	0.12	0.7	110	Tr	Tr	Tr

No. 17-	Food	Retinol µg	Carotene µg	Vitamin D µg	Vitamin E mg	Thiamin mg	Riboflavin mg	Niacin mg	Trypt/60 mg	Vitamin B_6 mg	Vitamin B_{12} µg	Folate µg	Pantothenate mg	Biotin µg	Vitamin C mg
	Non chocolate confectionery *continued*														
120	**Toffees**, mixed	0	0	0	N	0	0	0	0.4	0	0	0	0	0	0
121	**Turkish delight**, with nuts	0	6	0	0.19	0.03	0.01	0.1	0.2	0	0	3	Tr	Tr	0
122	without nuts	0	0	0	0	0.13	N	N	N	N	N	N	N	N	0

SAVOURY SNACKS

This section includes data for savoury snacks. Most of the values in the fifth edition of *The Composition of Foods* have been updated and many new products have been included. This section of the book now contains values for 29 foods compared with 8 in the fifth edition.

The composition of many savoury snacks may change if the product size or the recipe has changed. Amounts of protein, fat, carbohydrate, energy and salt in individual products may vary. Users are advised to check the label for more specific information.

No. *17-*	Food	Description and main data sources	Water g	Total Nitrogen g	Protein g	Fat g	Carbo-hydrate g	Energy value kcal	Energy value kJ
	Savoury snacks								
123	**Breadsticks**	10 samples, 3 brands	3.5	1.92	11.2	8.4	72.5	392	1661
124	**Chevda/chevra/chewra**	Recipe	3.6	2.80	17.5	32.3	35.1	492	2054
125	**Corn snacks**	20 samples, 7 brands including Wotsits, Monster Munch and Nik-Naks	3.3	1.12	7.0	31.9	54.3	519	2168
126	**Corn and starch snacks**	20 samples, 6 brands including Skips	2.6	0.61	3.8	31.2	55.8	505	2112
127	**Maize and rice flour snacks**	20 samples of Frazzles and Bacon Streaks	2.6	1.12	7.0	20.9	62.5	450	1892
128	**Mixed cereal and potato flour snacks**	40 samples. Shaped and flavoured snacks based on maize, potato and wheat including cheese balls	4.0	1.14	7.1	23.1	56.7	449	1883
129	**Oriental mix**	Ref. Cutrufelli and Pehrsson (1991)	2.6	3.20	20.0	40.7	*33.1*	*545*	*2283*
130	**Popcorn**, candied	Recipe	2.6	0.33	2.1	20.0	77.6	480	2018
131	plain	Recipe	0.9	0.99	6.2	42.8	48.7	593	2468
132	**Pork scratchings**	19 samples, 4 brands	2.1	7.66	47.9	46.0	0.2	606	2520
133	**Potato crisps**	20 samples, 8 brands; mixed plain and flavoured	2.8	0.91	5.7	34.2	53.3	530	2215
134	crinkle-cut	20 samples, 3 brands; mixed plain and flavoured	2.6	0.90	5.6	35.8	53.9	547	2282
135	jacket	20 samples, 7 brands; mixed plain and flavoured	1.3	1.02	6.4	32.4	51.3	510	2128
136	low fat	20 samples of different brands; mixed plain and flavoured	1.1	1.06	6.6	21.5	63.5	458	1924
137	square	20 samples, 2 brands; mixed plain and flavoured	4.5	1.02	6.4	21.2	57.7	433	1816
138	thick-cut	20 samples, 2 brands; mixed plain and flavoured	3.4	1.15	7.2	28.1	58.0	499	2090
139	thick, crinkle-cut	20 samples, 5 brands; mixed plain and flavoured	2.1	0.98	6.1	30.3	55.9	507	2119

Carbohydrate fractions and fatty acids, g per 100g food
Cholesterol, mg per 100g food

No. 17-	Food	Starch g	Total sugars g	Individual sugars: Gluc g	Fruct g	Sucr g	Malt g	Lact g	Dietary fibre: Southgate method g	Englyst method g	Fatty acids: Satd g	Mono-unsatd g	Poly-unsatd g	Cholesterol mg
	Savoury snacks													
123	**Breadsticks**	67.5	5.0	0.8	0.5	0	3.7	0	N	2.8	5.9	1.3	0.9	0
124	**Chevda/chevra/chewra**	31.0	2.0	N	N	N	N	N	6.0	3.8	N	N	N	0
125	**Corn snacks**	49.7	4.6	0.3	0.1	0.7	0	3.5	N	1.0	11.8	12.9	5.8	0
126	**Corn and starch snacks**	51.6	4.2	0.5	0.2	3.5	0	Tr	N	3.5	13.6	12.5	3.8	0
127	**Maize and rice flour snacks**	60.9	1.6	0.9	Tr	0.7	0	0	N	2.1	6.3	9.0	4.1	0
128	**Mixed cereal and potato flour snacks**	55.3	1.4	0.2	0.2	1.0	0	Tr	N	2.3	5.5	5.7	5.5	0
129	**Oriental mix**	N	N	N	N	N	N	0	N	N	N	N	N	0
130	**Popcorn**, candied	15.5	62.1	Tr	Tr	62.1	0	0	N	N	2.0	6.8	9.2	18
131	plain	47.6	1.1	0.1	0.1	0.9	0	0	N	N	4.3	14.5	19.7	0
132	**Pork scratchings**	Tr	0.2	0.2	Tr	Tr	0	0	N	0.3	N	N	N	N
133	**Potato crisps**	52.6	0.7	0.1	0	0.5	0	Tr	10.7	5.3	14.0	13.7	5.0	0
134	crinkle-cut	53.0	0.9	0.3	Tr	0.6	0	Tr	(11.5)	(5.7)	(14.6)	(14.3)	(5.2)	0
135	jacket	50.6	0.7	0.1	Tr	0.6	0	Tr	(9.6)	4.8	(13.3)	(13.0)	(4.7)	0
136	low fat	62.0	1.5	0.2	Tr	0.8	0	0.5	13.7	5.9	9.3	8.7	2.5	0
137	square	55.8	1.9	0.2	0.3	0.7	0	0.7	(9.3)	4.6	(8.7)	(8.5)	(3.1)	0
138	thick-cut	56.5	1.5	0.4	0.2	0.9	0	Tr	N	N	(11.5)	(11.3)	(4.1)	0
139	thick, crinkle-cut	55.4	0.5	0.1	Tr	0.4	0	Tr	(8.3)	4.1	(12.4)	(12.1)	(4.4)	0

Inorganic constituents per 100g food

No. 17-	Food	Na (mg)	K (mg)	Ca (mg)	Mg (mg)	P (mg)	Fe (mg)	Cu (mg)	Zn (mg)	Cl (mg)	Mn (mg)	Se (µg)	I (µg)
	Savoury snacks												
123	**Breadsticks**	860	160	26	25	110	1.2	0.12	0.7	630	0.48	N	N
124	**Chevda/chevra/chewra**	1000	540	53	110	250	5.1	0.29	2.5	1510	1.32	N	N
125	**Corn snacks**	1130	200	68	18	140	0.8	0.04	0.5	1840	0.13	(3)	N
126	**Corn and starch snacks**	1320	130	30	18	82	1.3	0.05	0.4	2060	0.13	(3)	N
127	**Maize and rice flour snacks**	1740	140	13	65	120	0.6	0.10	0.9	3040	0.34	(3)	N
128	**Mixed cereal and potato flour snacks**	1820	340	33	25	110	1.1	0.10	0.6	3820	0.29	N	N
129	**Oriental mix**	830	520	77	140	400	2.8	1.00	4.6	N	1.50	N	N
130	**Popcorn**, candied	56	75	6	26	58	0.4	N	0.7	100	0.10	N	3
131	plain	4	220	10	81	170	1.1	N	1.7	8	0.32	N	2
132	**Pork scratchings**	1320	300	32	18	180	2.4	0.20	1.6	2090	0.09	N	N
133	**Potato crisps**	840[a]	940	29	39	110	1.5	0.15	0.5	1310	0.27	N	N
134	crinkle-cut	770	1040	31	44	110	1.7	0.18	0.7	(1190)	0.28	N	N
135	jacket	520	1290	46	49	140	2.0	0.19	0.9	1040	0.42	N	N
136	low fat	730	1020	36	48	130	1.8	0.38	0.9	1200	0.37	N	N
137	square	1470	1310	92	49	160	0.3	0.24	0.7	(2270)	0.28	N	N
138	thick-cut	530	1440	38	62	120	1.9	0.24	1.0	(820)	0.39	N	N
139	thick, crinkle-cut	740	920	34	42	130	1.5	0.17	0.6	1280	0.32	N	N

[a] Na content ranged from 600mg to 1500mg per 100g. Lightly salted crisps contain about 400mg Na per 100g, and unsalted crisps a trace

No. 17-	Food	Retinol µg	Carotene µg	Vitamin D µg	Vitamin E mg	Thiamin mg	Riboflavin mg	Niacin mg	Trypt/60 mg	Vitamin B_6 mg	Vitamin B_{12} µg	Folate µg	Pantothenate mg	Biotin µg	Vitamin C mg
	Savoury snacks														
123	**Breadsticks**	0	Tr	0	0.44	0.12	0.08	1.6	3.9	0.10	Tr	18	0.60	2	0
124	**Chevda/chevra/chewra**	0	440	0	6.31	0.39	0.09	6.3	3.5	0.26	0	30	1.10	10	2
125	**Corn snacks**	0	460	0	5.80	0.19	0.16	0.9	0.7	0.13	0	49	N	N	Tr
126	**Corn and starch snacks**	0	145	0	N	0.27	0.10	1.2	0.4	0.11	0	35	N	N	Tr
127	**Maize and rice flour snacks**	0	0	0	N	0.38	0.15	2.1	1.1	0.17	0	40	N	N	Tr
128	**Mixed cereal and potato flour snacks**	0	0	0	N	0.20	0.13	2.6	1.2	0.12	Tr	24	N	N	Tr
129	**Oriental mix**	0	N	0	N	0.29	0.16	10.5	3.7	0.18	0	87	0.92	Tr	1
130	**Popcorn**, candied	52	98	0.1	3.75	0.06	0.04	0.3	0.2	0.07	0	3	0.10	1	0
131	plain	0	230	0	11.03	0.18	0.11	1.0	0.7	0.20	0	9	0.30	4	0
132	**Pork scratchings**	0	0	Tr	N	0.56	0.20	4.2	2.5	0.05	N	N	N	N	0
133	**Potato crisps**	0	2	0	5.83	0.21	0.08	3.2	1.3	0.81	0	38	0.93	N	6
134	crinkle-cut	0	(2)	0	(6.10)	(0.21)	(0.08)	(3.1)	1.4	(0.80)	Tr	(37)	(0.91)	N	Tr
135	jacket	0	(2)	0	(5.50)	0.21	0.12	5.1	1.5	0.49	0	70	(1.04)	N	19
136	low fat	0	(2)	0	3.47	0.19	0.14	5.0	1.6	0.46	0	48	N	N	14
137	square	0	(2)	0	N	N	N	N	1.5	N	0	N	N	N	4
138	thick-cut	0	(2)	0	(4.80)	(0.27)	(0.10)	(4.0)	1.7	(1.02)	0	(48)	(1.17)	N	Tr
139	thick, crinkle-cut	0	(2)	0	(5.20)	0.32	0.10	4.5	1.5	0.57	0	47	(0.99)	N	8

Composition of food per 100g

No. 17-	Food	Description and main data sources	Water g	Total Nitrogen g	Protein g	Fat g	Carbo-hydrate g	Energy value kcal	Energy value kJ
	***Savoury snacks** continued*								
140	**Potato and corn sticks**	20 samples, 9 brands; mixed plain and flavoured e.g. chipsticks, crunchy sticks	3.2	0.93	5.8	23.6	60.4	462	1938
141	**Potato and tapioca snacks**	20 samples, 3 brands; assorted types e.g. Waffles, Bitza Pizza, Wickettes	3.2	0.62	3.9	24.8	64.6	481	2018
142	**Potato rings**	18 samples, 3 brands; assorted flavours; Hula Hoop type	2.8	0.62	3.9	32.0	58.5	523	2186
143	**Pot savouries**	6 samples including assorted flavours of noodles, rice and chilli	8.9	1.86	11.6	10.9	58.8[a]	365	1541
144	*made up*	85g product made up with 215g water	74.2	0.53	3.3	3.1	16.7[b]	103	437
145	**Pretzels**	Ref. Cutrufelli and Pehrsson (1991)	3.3	1.46	9.1	3.5	*79.2*	*381*	*1596*
146	**Punjabi puri**	20 samples, 3 brands; assorted flavours	2.0	1.17	7.3	35.1	51.1	537	2240
147	**Puffed potato products**	20 samples, 3 brands; assorted flavours e.g. Quavers, Snaps, Chinese style crackers	3.1	0.35	2.2	33.0	56.2	517	2158
148	**Sev/ganthia**	Recipe, savoury Indian snack	3.8	2.88	18.0	25.3	41.1	454	1900
149	**Tortilla chips**	20 samples, 6 brands, maize chips	0.9	1.22	7.6	22.6	60.1	459	1927
150	**Twiglets**	20 samples, savoury wholewheat sticks	3.2	1.98	11.3	11.7	62.0	383	1617
151	**Wheat crunchies**	20 samples, 3 brands; assorted flavours	3.8	1.81	10.3	20.7	59.4	450	1891

[a] Includes 3.7g maltodextrins

[b] Includes 1.1g maltodextrins

Carbohydrate fractions and fatty acids, g per 100g food
Cholesterol, mg per 100g food

No.	Food	Starch	Total sugars	Individual sugars					Dietary fibre		Fatty acids			Cholest-erol
17-				Gluc	Fruct	Sucr	Malt	Lact	Southgate method	Englyst method	Satd	Mono-unsatd	Poly-unsatd	
		g	g	g	g	g	g	g	g	g	g	g	g	mg
	***Savoury snacks** continued*													
140	**Potato and corn sticks**	58.9	1.5	0.2	0.3	1.0	0	Tr	N	3.1	6.6	10.5	5.5	0
141	**Potato and tapioca snacks**	61.8	2.8	0.4	0.4	2.0	0	0	N	2.6	5.5	10.7	7.6	0
142	**Potato rings**	58.0	0.5	Tr	Tr	0.4	0	Tr	N	2.6	13.9	12.7	4.0	0
143	**Pot savouries**	46.9	8.2	1.3	1.9	3.9	0.8	0.3	N	N	N	N	N	0
144	*made up*	13.3	2.3	0.4	0.5	1.1	0.2	0.1	N	N	N	N	N	0
145	**Pretzels**	N	N	N	N	N	N	0	N	N	0.8	1.4	1.2	0
146	**Punjabi puri**	48.4	2.7	0.5	0.1	0.4	0	1.5	N	2.5	8.3	13.9	11.4	0
147	**Puffed potato products**	54.0	2.2	0.3	0.1	0.7	0	1.1	N	1.5	9.7	16.8	5.1	0
148	**Sev/ganthia**	40.3	0.8	0	0	0.8	0	0	11.1	8.8	2.6	8.2	12.1	0
149	**Tortilla chips**	58.9	1.2	0.1	0.1	1.0	0	0	N	(6.0)	4.0	10.6	6.7	0
150	**Twiglets**	60.9	1.1	Tr	Tr	1.1	0	Tr	N	10.3	4.9	4.4	1.8	0
151	**Wheat crunchies**	57.3	2.1	0.6	0.1	1.4	0	Tr	N	3.3	9.1	8.1	2.6	0

Inorganic constituents per 100g food

No.	Food	mg										µg	
17-		Na	K	Ca	Mg	P	Fe	Cu	Zn	Cl	Mn	Se	I
	Savoury snacks *continued*												
140	**Potato and corn sticks**	1090	840	20	37	120	1.2	0.18	0.6	1430	0.24	(1)	N
141	**Potato and tapioca snacks**	1350	530	40	29	94	1.8	0.12	0.6	2320	0.32	(1)	N
142	**Potato rings**	1070	540	22	28	100	1.0	0.16	0.7	(1650)	0.21	(1)	N
143	**Pot savouries**	1310	640	180	76	210	4.1	0.36	1.4	210	1.03	N	N
144	*made up*	370	180	51	22	59	1.2	0.10	0.4	61	0.29	N	N
145	**Pretzels**	1720	150	36	35	110	4.3	0.26	0.9	N	1.80	N	N
146	**Punjabi puri**	650	210	120	43	160	2.1	0.10	1.1	1240	0.56	N	N
147	**Puffed potato products**	1480	190	34	14	84	1.0	0.04	0.2	2630	0.14	(1)	N
148	**Sev/ganthia**	610	660	160	110	310	7.6	0.57	2.9	950	1.89	N	N
149	**Tortilla chips**	860	220	150	89	240	1.6	0.09	1.2	1400	0.43	N	N
150	**Twiglets**	1340	460	45	81	370	2.9	0.32	2.0	2520	1.61	N	N
151	**Wheat crunchies**	1180	260	38	53	220	2.2	0.19	1.3	2180	1.10	(4)	N

No. 17-	Food	Retinol µg	Carotene µg	Vitamin D µg	Vitamin E mg	Thiamin mg	Riboflavin mg	Niacin mg	Trypt/60 mg	Vitamin B_6 mg	Vitamin B_{12} µg	Folate µg	Pantothenate mg	Biotin µg	Vitamin C mg
	***Savoury snacks** continued*														
140	**Potato and corn sticks**	0	0	0	N	Tr	0.08	2.7	1.4	0.52	0	46	N	N	Tr
141	**Potato and tapioca snacks**	0	78	0	N	0.59	0.09	2.2	0.9	0.36	Tr	N	N	N	Tr
142	**Potato rings**	0	0	0	N	N	N	N	0.1	N	0	N	N	N	3
143	**Pot savouries**	0	N	0	N	N	N	N	N	N	0	N	N	N	0
144	*made up*	0	N	0	N	N	N	N	N	N	0	N	N	N	0
145	**Pretzels**	0	0	0	N	0.46	0.62	5.3	2.7	0.12	0	N	0.29	N	0
146	**Punjabi puri**	0	Tr	0	5.52	0.68	0.18	1.4	1.2	0.20	Tr	24	N	N	3
147	**Puffed potato products**	0	0	0	N	0.21	0.10	0.8	0.5	0.08	Tr	34	N	N	Tr
148	**Sev/ganthia**	0	170	0	N	0.29	0.09	0.9	2.2	0.25	0	74	0.90	N	1
149	**Tortilla chips**	0	455	0	1.94	0.17	0.09	1.8	0.8	0.31	0	19	N	N	Tr
150	**Twiglets**	0	Tr	0	2.47	0.37	0.48	7.8	2.3	0.38	0	78	1.54	15	Tr
151	**Wheat crunchies**	0	0	0	2.68	0.50	0.10	5.3	2.1	0.25	0	25	N	N	Tr

BEVERAGES

This section includes data for coffee, tea, carbonated and other soft drinks, fruit drinks, squashes and cordials. Milk drinks and fruit *juices* have not been included in this section but are covered in the *Milk and Milk Products* and the *Fruit and Nuts* supplements. Alcoholic beverages are given on pages 74 to pages 87 and distilled water on page 124.

Where appropriate, beverages have been made up with distilled water and the proportions given in the product description. Tap water can vary in composition by both area and source of supply, so users may wish to contact their local water board for the composition of tap water in their specific area. Hard waters may contain as much as 160mg calcium and 50mg magnesium per litre.

The vitamin composition of carbonated and soft drinks may vary from the values in these tables if manufacturers have added to or changed the fortification of products. Where possible, the tables give the values found during recent analyses. The vitamin C levels of many fruit drinks can also vary widely depending on the amount added, so the user should check the label to establish the vitamin C content of these drinks, although the levels fall quite rapidly after the container has been opened. Energy values do not include any contribution from citric or other organic acids.

The nutrients are given per 100g but many beverages are sold or measured by volume. The following table gives typical specific gravities of selected beverages. To convert the amounts per 100g into amounts per 100ml, multiply by the appropriate specific gravity.

Carbonated drinks	
Lucozade	1.070
Cola	1.040
Fruit juice drinks	1.040
Lemonade	1.020

Squashes and Cordials	
Blackcurrant fruit drinks, concentrated	1.280
Blackcurrant fruit juice drinks, ready to drink	1.050
Barley water, concentrated	1.100
Fruit drinks, concentrated	1.090-1.120
Fruit drinks, ready to drink	1.030-1.040
Fruit drinks, low calorie, concentrated	1.010-1.030
Fruit drinks, low calorie, ready to drink	1.010
Fruit juice drinks, ready to drink	1.030-1.040
High juice drinks, concentrated	1.150
High juice drinks, ready to drink	1.040
Lime juice cordial, concentrated	1.102

Composition of food per 100g

No. 17-	Food	Description and main data sources	Water g	Total Nitrogen g	Protein g	Fat g	Carbo-hydrate g	Energy value kcal	Energy value kJ
	Coffee								
152	**Coffee**, infusion, average	Average of strong and weak infusions	98.3	0.03	0.2	Tr	0.3	2	8
153	-, strong	Ground and filter coffee; 69g per litre water boiled in percolator and strained	97.8	0.03	0.2	Tr	0.3	2	8
154	-, weak	Ground and filter coffee; 34g per litre water boiled in percolator and strained	98.8	0.03	0.2	Tr	0.3	2	8
155	-, average, with single cream	225ml coffee infusion with 15g single cream	96.8	0.05	0.4	1.2	0.3	14	56
156	-, with whole milk	225ml coffee infusion with 28g whole milk	97.1	0.08	0.5	0.4	0.5	7	31
157	-, with semi-skimmed milk	225ml coffee infusion with 35g semi-skimmed milk	97.2	0.10	0.6	0.2	0.7	7	29
158	instant	10 jars, 2 brands	3.4	3.26[a]	14.6[b]	Tr	4.5	75	320
159	*made up*	Calculated from 2g instant coffee to 225ml water	99.2	0.02	0[c]	Tr[c]	Tr[c]	Tr[c]	2[c]
160	-, with whole milk	Calculated from 2g instant coffee to 225ml water with 28g whole milk	97.9	0.08	0.5	0.4	0.6	8	34
161	-, with semi-skimmed milk	Calculated from 2g instant coffee to 225ml water with 35g semi-skimmed milk	97.9	0.09	0.6	0.2	0.7	7	29
162	**Coffee and chicory essence**	7 bottles of the same brand (CAMP)	36.9	0.33[d]	1.6[b]	0.2	56.0	218	931
163	with water	10g essence with 225ml water	97.3	0.01	0.1	Tr	2.4	9	40
164	**Coffee**, Irish	Recipe	(85.1)	0.06	0.4	7.8	3.2	84[e]	346[e]

[a] Includes 0.93g purine nitrogen.
[b] (Total N - purine N) x 6.25
[c] Instant coffee with single cream provides 0.3g protein, 1.2g fat, 0.3g carbohydrate, 13 kcal and 54 kJ per 100g (2g coffee with 225ml water and 15g single cream)
[d] Includes 0.08g purine nitrogen
[e] Includes contribution from 5.1g alcohol

Carbohydrate fractions and fatty acids, g per 100g food
Cholesterol, mg per 100g food

No. 17-	Food	Starch g	Total sugars g	Individual sugars Gluc g	Fruct g	Sucr g	Malt g	Lact g	Dietary fibre Southgate method g	Englyst method g	Fatty acids Satd g	Mono-unsatd g	Poly-unsatd g	Cholesterol mg
	Coffee													
152	**Coffee**, infusion, average	0	0	0	0	0	0	0	0	0	Tr	Tr	Tr	0
153	-, strong	0	0	0	0	0	0	0	0	0	Tr	Tr	Tr	0
154	-, weak	0	0	0	0	0	0	0	0	0	Tr	Tr	Tr	0
155	-, average with single cream	0	0.3	0	0	0	0	0.3	0	0	0.7	0.3	Tr	3
156	-, with whole milk	0	0.5	0	0	0	0	0.5	0	0	0.3	0.1	Tr	2
157	-, with semi-skimmed milk	0	0.7	0	0	0	0	0.7	0	0	0.1	0.1	Tr	1
158	instant	4.5	0	0	0	0	0	0	0	0	Tr	Tr	Tr	0
159	*made up*	Tr	0	0	0	0	0	0	0	0	0	0	0	0
160	-, with whole milk	Tr	0.5	0	0	0	0	0.5	0	0	0.3	0.1	Tr	2
161	-, with semi-skimmed milk	Tr	0.7	0	0	0	0	0.7	0	0	0.1	0.1	Tr	1
162	**Coffee and chicory essence**	2.2	53.8	2.9	3.4	47.5	0	0	0	0	Tr	Tr	Tr	0
163	with water	0.1	2.3	0.1	0.1	2.0	0	0	0	0	0	0	0	0
164	**Coffee**, Irish	0	3.5	0	0	2.8	0	0.4	0	0	4.9	2.3	0.2	21

No. 17-	Food	mg										µg	
		Na	K	Ca	Mg	P	Fe	Cu	Zn	Cl	Mn	Se	I
	Coffee												
152	**Coffee**, infusion, average	Tr	92	3	8	7	0.1	Tr	Tr	3	0.05	Tr	Tr
153	-, strong	1	120	3	10	8	0.1	0.01	0.1	3	0.07	Tr	Tr
154	-, weak	Tr	64	2	5	5	Tr	Tr	Tr	2	0.03	Tr	Tr
155	-, average, with single cream	3	94	9	8	11	0.1	Tr	0.1	8	0.05	Tr	Tr
156	-, with whole milk	6	97	15	8	16	0.1	Tr	0.1	14	0.04	Tr	2
157	-, with semi-skimmed milk	7	10	19	8	19	0.1	Tr	0.1	16	0.04	Tr	2
158	instant	81	3780	140	330	310	4.6	0.62	1.1	65	2.10	9	Tr
159	*made up*	Tr	33	1	3	3	Tr	Tr	Tr	1	0.02	Tr	Tr
160	-, with whole milk	7	45	14	4	13	Tr	Tr	0.1	11	Tr	Tr	2
161	-, with semi-skimmed milk	8	49	17	4	15	Tr	Tr	0.1	14	0.02	Tr	2
162	**Coffee and chicory essence**	65	750	30	39	90	0.7	0.60	N	85	N	N	N
163	with water	3	32	1	2	4	Tr	0.03	N	4	N	N	N
164	**Coffee**, Irish	6	72	11	6	13	0.1	Tr	0.1	11	0.03	Tr	Tr

No. 17-	Food	Retinol µg	Carotene µg	Vitamin D µg	Vitamin E mg	Thiamin mg	Riboflavin mg	Niacin mg	Trypt 60 mg	Vitamin B_6 mg	Vitamin B_{12} µg	Folate µg	Pantothenate mg	Biotin µg	Vitamin C mg
	Coffee														
152	**Coffee**, infusion, average	0	0	0	Tr	Tr	0.01	0.7	0	Tr	0	Tr	Tr	3	0
153	-, strong	0	0	0	Tr	Tr	0.01	0.9	0	Tr	0	Tr	Tr	3	0
154	-, weak	0	0	0	Tr	Tr	Tr	0.5	0	Tr	0	Tr	Tr	2	0
155	-, average, with single cream	20	8	Tr	0.03	Tr	0.02	0.6	Tr	Tr	Tr	1	0.02	3	0
156	-, with whole milk	6	2	Tr	0.01	Tr	0.03	0.6	0.1	0.01	Tr	1	0.04	3	0
157	-, with semi-skimmed milk	3	1	Tr	Tr	0.01	0.03	0.6	0.1	0.01	Tr	1	0.04	3	0
158	instant	0	N	0	Tr	0.04	0.21	24.8[a]	2.9	0.02	0	11	Tr	67	0
159	*made up*	0	Tr	0	0	0	Tr	0.2	0.1	0	0	Tr	0	1	0
160	-, with whole milk	6	2	Tr	0.01	Tr	0.02	0.2	0.1	Tr	Tr	1	0.04	1	0
161	-, with semi-skimmed milk	3	1	Tr	Tr	0.01	0.03	0.2	0.1	0.01	Tr	1	0.04	1	0
162	**Coffee and chicory essence**	0	N	0	N	0	0.03	2.8	N	N	0	N	N	N	0
163	with water	0	N	0	N	0	0	0.1	N	N	0	N	N	N	0
164	**Coffee**, Irish	97	53	Tr	0.18	Tr	0.03	0.4	0.1	Tr	Tr	1	0.03	2	0

[a] Can be as high as 39mg per 100g. Decaffeinated instant coffee contains about the same

Composition of food per 100g

No. 17-	Food	Description and main data sources	Water g	Total Nitrogen g	Protein g	Fat g	Carbo-hydrate g	Energy value kcal	Energy value kJ
Tea									
165	**Tea**, black, infusion, average	15g leaves per litre water, strained after 5 minutes	99.5	Tr	0.1	Tr	Tr	Tr	2
166	-, weak	10g leaves per litre water, strained after 5 minutes; 10 samples	99.7	Tr	0.1	Tr	Tr	Tr	2
167	-, strong	10g leaves per 500ml water, strained after 5 minutes; 10 samples	99.3	Tr	0.1	Tr	Tr	Tr	2
168	average, with whole milk	225ml infusion with 28g whole milk	98.2	0.06	0.4	0.4	0.5	8	32
169	-, with semi-skimmed milk	225ml infusion with 35g semi-skimmed milk	98.2	0.08	0.5	0.2	0.7	7	28
170	Chinese, leaves, infusion	Analysis and literature sources	99.6	0.02	0.1	0	*0.2*	*1*	*5*
171	green, infusion	Literature sources; Sencha Fuku Jyu and Banch type	99.7	0.02	0.1	0	Tr	Tr	Tr
172	herbal, infusion	Ref. Cutrufelli and Matthews (1986)	99.7	0	0	Tr	*0.2*	*1*	*13*
173	instant powder	8 samples of the same brand (Lift)	8.3	Tr	Tr	Tr	81.7	306	1307
174	-, with water	Calculated from 6g powder to 225ml water	97.6	0	0	0	2.1	8	34
Carbonated drinks									
175	**Cola**	10 samples, 6 brands	89.7	Tr	Tr	0	10.3	39[a]	165[a]
176	**Dr Pepper**	Ref. Cutrufelli and Matthews (1986)	89.4	0	0	0.1	*10.4*	*41*	*171*
177	**Fruit juice drink**, carbonated, ready to drink	Mixed sample of different brands; bottles and cans; orange, lemon, apple and tropical fruit flavours e.g. Citrus Spring, Fanta, Orangina and Tango	89.7	Tr	Tr	Tr	10.3	39	165

[a] Low calorie cola provides 0.5 kcals and 2 kJ

Carbohydrate fractions and fatty acids, g per 100g food
Cholesterol, mg per 100g food

No. 17-	Food	Starch	Total sugars	Individual sugars					Dietary fibre		Fatty acids			Cholesterol
				Gluc	Fruct	Sucr	Malt	Lact	Southgate method	Englyst method	Satd	Mono-unsatd	Poly-unsatd	
		g	g	g	g	g	g	g	g	g	g	g	g	mg
	Tea													
165	**Tea**, black, infusion, average	0	Tr	0	0	0	0	0	0	0	Tr	Tr	Tr	0
166	-, weak	0	Tr	0	0	0	0	0	0	0	Tr	Tr	Tr	0
167	-, strong	0	Tr	0	0	0	0	0	0	0	Tr	Tr	Tr	0
168	average, with whole milk	0	0.5	0	0	0	0	0.5	0	0	0.3	0.1	Tr	2
169	-, with semi-skimmed milk	0	0.7	0	0	0	0	0.7	0	0	0.1	0.1	0	1
170	Chinese, leaves, infusion	0	Tr	0	0	0	0	0	0	0	0	0	0	0
171	green, infusion	0	Tr	0	0	0	0	0	0	0	0	0	0	0
172	herbal, infusion	0	0	0	0	0	0	0	0	0	Tr	Tr	Tr	0
173	instant powder	0	81.7	81.7	0	0	0	0	N	N	0	0	0	0
174	-, with water	0	2.1	2.1	0	0	0	0	N	N	0	0	0	0
	Carbonated drinks													
175	**Cola**	Tr	10.9	3.5	3.4	4.0	0	0	0	0	0	0	0	0
176	**Dr Pepper**	0	N	N	N	N	N	0	0	0	0	0	0	0
177	**Fruit juice drink**, carbonated, ready to drink	0	10.3	2.2	2.1	5.9	0.1	0	Tr	Tr	Tr	Tr	Tr	0

No. 17-	Food	mg										µg	
		Na	K	Ca	Mg	P	Fe	Cu	Zn	Cl	Mn	Se	I
	Tea												
165	**Tea**, black, infusion, average	Tr	27	Tr	2	2	Tr	0.01	Tr	1	0.15	Tr	Tr
166	-, weak	Tr	18	Tr	1	1	Tr	0.01	Tr	1	0.10	Tr	Tr
167	-, strong	Tr	35	Tr	2	3	Tr	0.01	Tr	1	0.19	Tr	Tr
168	average, with whole milk	6	40	13	3	12	0	0.01	0.1	12	0.13	Tr	2
169	-, with semi-skimmed milk	7	44	16	3	15	0	0.01	0.1	14	0.13	Tr	2
170	Chinese, leaves, infusion	Tr	18	2	1	1	0.1	Tr	Tr	Tr	Tr	Tr	Tr
171	green, infusion	1	20	2	Tr	1	0.1	Tr	Tr	Tr	Tr	Tr	Tr
172	herbal, infusion	Tr	9	2	1	Tr	0.1	0.02	Tr	Tr	0.04	Tr	Tr
173	instant powder	200	170	2	7	11	0.1	Tr	Tr	59	4.70	Tr	Tr
174	-, with water	5	4	Tr	Tr	Tr	Tr	0	0	2	0.12	Tr	Tr
	Carbonated drinks												
175	**Cola**	5	1	6	1	30	Tr	Tr	Tr	Tr	Tr	Tr	Tr
176	**Dr Pepper**	10	1	3	0	11	Tr	Tr	Tr	N	Tr	Tr	Tr
177	**Fruit juice drink**, carbonated, ready to drink	8	27	7	7	2	Tr	Tr	Tr	3	Tr	Tr	Tr

No. 17-	Food	Retinol µg	Carotene µg	Vitamin D µg	Vitamin E mg	Thiamin mg	Riboflavin mg	Niacin mg	Trypt/60 mg	Vitamin B_6 mg	Vitamin B_{12} µg	Folate µg	Pantothenate mg	Biotin µg	Vitamin C mg
	Tea														
165	**Tea**, black, infusion, average	0	0	0	N	Tr	0.02	0	0	Tr	0	3	0.04	1	0
166	-, weak	0	0	0	N	Tr	0.01	0	0	Tr	0	1	0.03	Tr	0
167	-, strong	0	0	0	N	Tr	0.02	0	0	0.01	0	5	0.05	1	0
168	average, with whole milk	6	2	0	0.01	0	0.03	0	0.1	0.01	Tr	3	0.07	1	0
169	-, with semi-skimmed milk	3	1	0	0	0.01	0.04	0	0.1	0.01	Tr	3	0.08	1	0
170	Chinese, leaves, infusion	0	0	0	N	Tr	Tr	Tr	0	Tr	0	N	Tr	Tr	0
171	green, infusion	0	0	0	N	0	0.02	0.1	0	Tr	0	Tr	Tr	Tr	3
172	herbal, infusion	0	0	0	N	0.01	Tr	0	0	0	0	1	0.01	0	0
173	instant powder	0	Tr	0	N	0.01	0.04	0.3	0	0.02	0	4	0.13	1	175
174	-, with water	0	0	0	Tr	Tr	Tr	Tr	0	Tr	0	Tr	Tr	Tr	5
	Carbonated drinks														
175	**Cola**	0	0	0	0	0	0	0	0	0	0	0	0	0	0
176	**Dr Pepper**	0	0	0	0	0	0	0	0	0	0	0	0	0	0
177	**Fruit juice drink**, carbonated, ready to drink	0	94	0	Tr	Tr	Tr	Tr	Tr	Tr	0	1	Tr	Tr	1[a]

[a] Fortified product contains 14mg vitamin C per 100g

Composition of food per 100g

No. 17-	Food	Description and main data sources	Water g	Total Nitrogen g	Protein g	Fat g	Carbo-hydrate g	Energy value kcal	Energy value kJ
	Carbonated drinks *continued*								
178	**Ginger ale**, dry	10 samples, 5 brands	95.9	0	0	0	3.9	15	62
179	**Lemonade**	10 samples, 8 brands	93.8	Tr	Tr	0	5.8	22	93
180	**Lucozade**	10 samples including lemon, orange, tropical flavours	81.8	0.01	Tr	0	16.0[a]	60	256
181	**Root beer**	Ref. Cutrufelli and Matthews (1986)	89.3	0	0	0	*10.6*	*41*	*172*
182	**Soda**, club	Ref. Cutrufelli and Matthews (1986)	99.9	0	0	0	0	0	0
183	cream	Ref. Cutrufelli and Matthews (1986)	86.7	0	0	0	*13.3*	*51*	*215*
184	**Tonic water**	Ref. Cutrufelli and Matthews (1986)	91.1	0	0	0	*8.8*	*33*	*141*
	Squash and cordials								
185	**Barley water**, concentrated	Mixed sample and manufacturers' information; orange and lemon flavours	75.2	0.05	0.3	Tr	*18.5*	*71*	*301*
186	*made up*	Calculated from 50g concentrate to 200ml water	95.0	0.02	0.1	0	*3.7*	*14*	*60*
187	**Blackcurrant juice drink**, concentrated	Mixed sample and manufacturer's data (Ribena and own brands)	40.9	0.02	0.1	0	60.8[b]	228	975
188	*made up*	Calculated from 38g concentrate to 250ml water	92.2	Tr	Tr	0	8.0[b]	30	129
189	**Fruit drink/squash**, concentrated	Mixed sample; citrus, apple and mixed fruit flavours	74.4	0.02	0.1	Tr	24.8[c]	93	399
190	*made up*	Calculated from 50g concentrate to 200ml water	94.9	Tr	Tr	0	5.0[b]	19	80

[a] Includes 1.7g oligosaccharides
[b] Includes oligosaccharides
[c] Includes 0.2g oligosaccharides

Carbohydrate fractions and fatty acids, g per 100g food
Cholesterol, mg per 100g food

No. 17-	Food	Starch g	Total sugars g	Individual sugars: Gluc g	Fruct g	Sucr g	Malt g	Lact g	Dietary fibre: Southgate method g	Englyst method g	Fatty acids: Satd g	Mono-unsatd g	Poly-unsatd g	Cholesterol mg
	***Carbonated drinks** continued*													
178	**Ginger ale**, dry	0	3.9	1.7	1.6	0.5	0	0	0	0	0	0	0	0
179	**Lemonade**	0	5.8	1.5	1.4	2.8	0.1	0	0	0	0	0	0	0
180	**Lucozade**	Tr	14.3	7.5	4.6	0.2	2.0	0	0	0	0	0	0	0
181	**Root beer**	0	(10.6)	N	N	N	0	0	0	0	0	0	0	0
182	**Soda**, club	0	0	0	0	0	0	0	0	0	0	0	0	0
183	cream	0	N	N	N	N	0	0	0	0	0	0	0	0
184	**Tonic water**	0	N	N	N	N	N	0	0	0	0	0	0	0
	Squash and cordials													
185	**Barley water**, concentrated	0	(18.5)	N	N	N	N	0	Tr	Tr	Tr	Tr	Tr	0
186	*made up*	0	(3.7)	N	N	N	N	0	0	0	0	0	0	0
187	**Blackcurrant juice drink**, concentrated	Tr	59.1	9.9	8.8	38.9	1.4	0	0	0	0	0	0	0
188	*made up*	0	7.8[a]	1.3	1.2	5.1	0.2	0	0	0	0	0	0	0
189	**Fruit drink/squash**, concentrated	0	24.6	10.3	10.3	3.1	0.9	0	Tr	Tr	Tr	Tr	Tr	0
190	*made up*	0	4.9	2.1	2.1	0.6	0.2	0	0	0	0	0	0	0

[a] Reduced sugar version contains approximately 5.1g sugars per 100g

Inorganic constituents per 100g food

No. 17-	Food	mg										µg	
		Na	K	Ca	Mg	P	Fe	Cu	Zn	Cl	Mn	Se	I
	***Carbonated drinks** continued*												
178	**Ginger ale**, dry	N	N	N	N	N	N	N	N	N	Tr	Tr	Tr
179	**Lemonade**	7	15	5	1	Tr	Tr	Tr	Tr	2	Tr	Tr	Tr
180	**Lucozade**	8	27	7	7	2	Tr	Tr	Tr	3	Tr	Tr	Tr
181	**Root beer**	13	1	5	1	0	0.1	0.01	0.1	N	Tr	Tr	Tr
182	**Soda**, club	21	2	5	1	0	N	N	0.1	N	Tr	Tr	Tr
183	cream	12	1	5	1	0	0.1	Tr	Tr	N	Tr	Tr	Tr
184	**Tonic water**	4	0	1	0	0	Tr	Tr	Tr	Tr	Tr	Tr	Tr
	Squash and cordials												
185	**Barley water**, concentrated	15	27	5	2	4	Tr	Tr	Tr	3	Tr	Tr	Tr
186	*made up*	3	5	1	Tr	1	0	0	0	1	0	0	0
187	**Blackcurrant juice drink**, concentrated	16	92	8	2	3	0.2	0.01	0.1	2	Tr	Tr	Tr
188	*made up*	2	12	1	Tr	Tr	Tr	Tr	Tr	Tr	0	0	0
189	**Fruit drink/squash**, concentrated	40	27	6	1	2	Tr	Tr	Tr	4	Tr	Tr	Tr
190	*made up*	8	5	1	Tr	Tr	0	0	0	1	0	0	0

No. 17-	Food	Retinol µg	Carotene µg	Vitamin D µg	Vitamin E mg	Thiamin mg	Riboflavin mg	Niacin mg	Trypt/60 mg	Vitamin B_6 mg	Vitamin B_{12} µg	Folate µg	Pantothenate mg	Biotin µg	Vitamin C mg
	Carbonated drinks *continued*														
178	**Ginger ale**, dry	0	0	0	0	0	0	0	0	0	0	0	0	0	0
179	**Lemonade**	0	Tr	0	Tr	Tr	Tr	Tr	Tr	Tr	0	Tr	Tr	Tr	Tr[a]
180	**Lucozade**	0	835	0	0	Tr	Tr	Tr	Tr	Tr	0	1	Tr	Tr	8
181	**Root beer**	0	0	0	0	0	0	0	0	0	0	0	0	0	0
182	**Soda**, club	0	0	0	0	0	0	0	0	0	0	0	0	0	0
183	cream	0	0	0	0	0	0	0	0	0	0	0	0	0	0
184	**Tonic water**	0	0	0	0	0	0	0	0	0	0	0	0	0	0
	Squash and cordials														
185	**Barley water**, concentrated	0	630	0	N	Tr	Tr	0.1	Tr	0.01	0	5	Tr	Tr	10
186	*made up*	0	125	0	N	0	0	Tr	0	0	0	1	0	0	2
187	**Blackcurrant juice drink**, concentrated	0	N	0	N	Tr	Tr	7.8[b]	Tr	1.01[c]	3[c]	Tr	Tr	Tr	78[b]
188	*made up*	0	N	0	N	0	0	1.0	0	0.13	0	0	0	0	10
189	**Fruit drink/squash**, concentrated	0	690	0	N	Tr	Tr	0.1	Tr	0.01	0	2	Tr	Tr	25[d]
190	*made up*	0	140	0	N	0	0	Tr	0	Tr	0	Tr	0	0	5

[a] 5-15mg vitamin C per 100g may be added to some brands
[b] When fresh, contains 9.5mg niacin and 107mg vitamin C per 100g
[c] These are declared amounts and represent levels present at the end of shelf life
[d] Unfortified product contains 2mg vitamin C per 100g

No. 17-	Food	Description and main data sources	Water g	Total Nitrogen g	Protein g	Fat g	Carbo-hydrate g	Energy value kcal	Energy value kJ
	Squash and cordials *continued*								
191	**Fruit drink**, low calorie, concentrated	10 samples, 2 brands; lemon, orange and mixed fruit flavours	97.3	0.02	0.1	Tr	0.8	3	15
192	*made up*	Calculated from 50g concentrate to 200ml water	99.5	Tr	Tr	0	0.2	1	3
193	low sugar, concentrated, fortified	10 samples of whole orange drink (Kia Ora)	92.5	0.03	0.2	Tr	5.1	20	85
194	*made up*	Calculated from 50g concentrate to 200ml water	98.5	0.01	Tr	0	1.0	4	17
195	**Fruit juice drink**, ready to drink	Mixed sample; lemon, orange, apple and mixed fruit flavours	89.5	0.02	0.1	Tr	9.8	37	159
196	low calorie, ready to drink	10 samples, 2 brands; mixed fruit flavours	96.6	0.03	0.2	Tr	2.5	10	43
197	**High juice drink**, concentrated	10 samples, 5 brands; orange and lemon flavours	64.8	0.04	0.3	Tr	33.0	125	533
198	*made up*	Calculated from 50g concentrate to 200ml water	93.0	0.01	0.1	0	6.6	25	107
199	**Lemonade**, homemade	Recipe	83.4	Tr	Tr	Tr	16.8	63	269
200	**Lime juice cordial**, concentrated	6 bottles of the same brand (Roses)	70.5	0.01	0.1	0	29.8	112	479
201	*made up*	Calculated from 50g concentrate to 200ml water	94.1	Tr	Tr	0	6.0	22	96
202	**Milkshake syrup**, concentrated	10 samples, 3 brands; assorted flavours	60.5	Tr	Tr	0	35.5	133	568
203	*made up with whole milk*	Calculated from 40g syrup to 250ml milk	84.0	0.43	2.8	3.4	9.0	75	315
204	*made up with semi-skimmed milk*	Calculated from 40g syrup to 250ml milk	85.8	0.45	2.9	1.4	9.3	59	250
205	**Rosehip syrup**, concentrated	9 bottles, 4 brands	32.5	Tr	Tr	0	61.9	232	990
206	*made up*	Calculated from 50g syrup to 200ml water	86.5	0	0	0	12.4	46	198

Carbohydrate fractions and fatty acids, g per 100g food
Cholesterol, mg per 100g food

No. 17-	Food	Starch g	Total sugars g	Individual sugars: Gluc g	Fruct g	Sucr g	Malt g	Lact g	Dietary fibre: Southgate method g	Englyst method g	Fatty acids: Satd g	Mono-unsatd g	Poly-unsatd g	Cholesterol mg
	Squash and cordials *continued*													
191	**Fruit drink**, low calorie, concentrated	0	0.8	0.3	0.4	0.1	Tr	0	Tr	Tr	Tr	Tr	Tr	0
192	*made up*	0	0.2	0.1	0.1	Tr	0	0	0	0	0	0	0	0
193	low sugar, concentrated, fortified	0	5.1	2.5	2.5	0	0.1	0	Tr	Tr	Tr	Tr	Tr	0
194	*made up*	0	1.0	0.5	0.5	0	Tr	0	0	0	0	0	0	0
195	**Fruit juice drink**, ready to drink	0	9.8	2.7	3.7	3.4	Tr	0	Tr	Tr	Tr	Tr	Tr	0
196	low calorie, ready to drink	0	2.5	0.8	0.9	0.8	Tr	0	Tr	Tr	Tr	Tr	Tr	0
197	**High juice drink**, concentrated	0	33.0	13.0	12.7	7.3	Tr	0	Tr	Tr	Tr	Tr	Tr	0
198	*made up*	0	6.6	2.6	2.5	1.5	0	0	0	0	0	0	0	0
199	**Lemonade**, homemade	0	16.8	Tr	Tr	16.7	0	0	Tr	Tr	Tr	Tr	Tr	0
200	**Lime juice cordial**, concentrated	Tr	29.8	11.5	11.0	5.9	1.4	0	0	0	0	0	0	0
201	*made up*	0	6.0	2.3	2.2	1.2	0.3	0	0	0	0	0	0	0
202	**Milkshake syrup**, concentrated	0	35.5	13.2	12.2	10.1	0	0	0	0	0	0	0	0
203	*made up with whole milk*	0	9.0	1.8	1.7	1.4	0	4.1	0	0	2.1	0.9	0.1	12
204	*made up with semi-skimmed milk*	0	9.3	1.8	1.7	1.4	0	4.3	0	0	0.9	0.4	Tr	6
205	**Rosehip syrup**, concentrated	0.1	61.8[a]	11.1	3.6	44.2	0	0	0	0	0	0	0	0
206	*made up*	Tr	12.4[b]	2.2	0.7	8.8	0	0	0	0	0	0	0	0

[a] Includes 2.9g of other sugars

[b] Includes 0.7g other sugars

Inorganic constituents per 100g food

No.	Food	mg										µg	
17-		Na	K	Ca	Mg	P	Fe	Cu	Zn	Cl	Mn	Se	I
	***Squash and cordials** continued*												
191	**Fruit drink**, low calorie, concentrated	40	31	5	1	2	Tr	Tr	Tr	3	Tr	Tr	Tr
192	*made up*	8	6	1	Tr	Tr	0	0	0	1	0	0	0
193	low sugar, concentrated, fortified	40	20	8	1	2	Tr	Tr	Tr	4	Tr	Tr	Tr
194	*made up*	8	4	2	Tr	Tr	0	0	0	1	0	0	0
195	**Fruit juice drink**, ready to drink	5	44	6	3	2	Tr	Tr	Tr	3	Tr	Tr	Tr
196	low calorie, ready to drink	5	48	5	3	3	Tr	Tr	Tr	2	0.03	Tr	Tr
197	**High juice drink**, concentrated	11	110	14	4	5	Tr	Tr	Tr	6	Tr	Tr	Tr
198	*made up*	2	23	3	1	1	0	0	0	1	0	0	0
199	**Lemonade**, homemade	1	10	(4)	(1)	(1)	(0)	(0.02)	Tr	(0)	Tr	Tr	N
200	**Lime juice cordial**, concentrated	8	49	9	4	5	0.3	0.07	N	4	Tr	Tr	Tr
201	*made up*	2	1	2	1	1	0.1	0.01	N	1	0	0	0
202	**Milkshake syrup**, concentrated	54	21	7	4	4	0.2	Tr	Tr	61	0.05	Tr	Tr
203	*made up with whole milk*	55	120	100	10	80	0.1	0	0.3	95	0.01	1	13
204	*made up with semi-skimmed milk*	55	140	110	10	85	0.1	0	0.3	100	0.01	(1)	(13)
205	**Rosehip syrup**, concentrated	280	26	N	N	N	N	N	N	N	N	Tr	Tr
206	*made up*	56	5	N	N	N	N	N	N	N	N	0	0

No. 17-	Food	Retinol µg	Carotene µg	Vitamin D µg	Vitamin E mg	Thiamin mg	Riboflavin mg	Niacin mg	Trypt/60 mg	Vitamin B_6 mg	Vitamin B_{12} µg	Folate µg	Pantothenate mg	Biotin µg	Vitamin C mg
	***Squash and cordials** continued*														
191	**Fruit drink** low calorie, concentrated	0	N	0	Tr	Tr	Tr	0.1	Tr	0.01	0	2	0.05	Tr	Tr
192	*made up*	0	N	0	0	0	0	Tr	0	Tr	0	Tr	0.01	0	0
193	low sugar, concentrated, fortified	0	N	0	Tr	Tr	Tr	0.1	Tr	0.01	0	3	Tr	Tr	59
194	*made up*	0	N	0	0	0	0	0	0	0	0	1	0	0	12
195	**Fruit juice drink**, ready to drink	0	N	0	N	Tr	Tr	0.1	Tr	0.01	0	2	Tr	Tr	23[a]
196	low calorie, ready to drink	0	Tr	0	Tr	0.02	Tr	0.1	Tr	0.01	0	2	0.06	Tr	5
197	**High juice drink**, concentrated	0	N	0	N	0.01	0.01	0.1	Tr	0.03	0	4	Tr	Tr	12
198	*made up*	0	N	0	N	0	0	Tr	0	0.01	0	1	0	0	2
199	**Lemonade** homemade	0	1	0	N	0	0	0	Tr	0.01	0	(1)	0.01	Tr	4
200	**Lime juice cordial**, concentrated	0	Tr	0	Tr	Tr	Tr	Tr	Tr	Tr	0	Tr	Tr	Tr	Tr
201	*made up*	0	0	0	0	0	0	0	0	0	0	0	0	0	0
202	**Milkshake syrup**, concentrated	0	Tr	0	Tr	Tr	0.01	0.1	Tr	Tr	0	Tr	0.03	Tr	0
203	*made up with whole milk*	45	18	0	0.08	0.03	0.15	0.1	0.6	0.05	0	5	0.31	2	1
204	*made up with semi-skimmed milk*	18	8	Tr	0.03	0.03	0.16	0.1	0.7	0.05	Tr	5	0.28	2	1
205	**Rosehip syrup**, concentrated	0	(500)	0	Tr	0	Tr	Tr	Tr	Tr	0	Tr	Tr	Tr	295
206	*made up*	0	(100)	0	0	0	0	0	0	0	0	0	0	0	59

[a] Unfortified product contains 7mg vitamin C per 100g

ALCOHOLIC BEVERAGES

Because alcoholic beverages are normally measured by volume, the data in this section (in contrast to all other sections of the book) are presented as amounts per 100ml. To determine the nutrients in 100 grams of the beverage, the values should be divided by the specific gravity. Typical specific gravities for the main types of alcoholic beverages are given in the table below.

The specific gravity varies with the composition of the drink. In general, it increases with the amount of solids (mainly sugars) and decreases with the amount of alcohol because the specific gravity of ethyl alcohol itself is only 0.79. Typical ranges of percentage of alcohol by volume of selected beers and lagers are given in the Appendix on page 168.

The water content of most of the alcoholic drinks has been calculated by difference.

Specific gravities of typical alcoholic beverages

Beers		**Ciders**	
Beer, bitter, canned	1.008	Cider, dry	1.007
draught	1.004	low alcohol	1.020
keg	1.001	sweet	1.012
low alcohol	1.020	vintage	1.017
Brown ale, bottled	1.008		
Lager, bottled	1.005	**Wines**	
Lager, low alcohol	1.010		
Mild, draught	1.009	Champagne	0.995
Pale ale, bottled	1.003	Red wine	0.998
Stout, bottled	1.014	Rosé wine, medium	1.003
Strong ale	1.018	White wine, dry	0.995
		medium	1.005
		sparkling	0.995
		sweet	1.016

Fortified wines

Port	1.026
Sherry, dry	0.998
medium	0.998
sweet	1.009
Vermouth, dry	1.005
sweet	1.046

Liqueurs

Advocaat	1.093
Cherry brandy	1.093
Curaçao	1.052

Spirits

40% volume	0.950

No. 17-	Food	Description and main data sources	Water g	Alcohol g	Total Nitrogen g	Protein g	Fat g	Carbo-hydrate g	Energy value kcal	Energy value kJ
	Beers									
207	**Bitter**	5 samples from different brewers; canned, draught and bottled	(93.9)	2.9	0.05	0.3	Tr	2.2	30	124
208	best/premium	Mixed sample from different brewers	(93.0)	3.4	0.05	0.3	Tr	2.2	33	139
209	low alcohol	10 samples from different brewers	94.7	0.6	0.03	0.2	0	2.1[a]	13	54
210	**Brown ale**, bottled	Mixed sample from different brewers	(93.3)	2.5	0.04	0.3	Tr	3.0	30	126
211	**Lager**	Mixed sample; Skol, Hofmeister, Tennents, Carling Black Label, Stella Artois and Fosters; canned and draught	(93.0)	4.0	0.05	0.3	Tr	Tr	29	121
212	alcohol-free	10 samples; Kaliber and Barbican	96.3	Tr	0.06	0.4	Tr	1.5[b]	7	31
213	low alcohol	10 samples; Carlton LA, Swan Light, Tennents LA	97.0	0.5	0.04	0.2	0	1.5[c]	10	41
214	premium	10 samples; Carlsberg Special Brew and Heldenbrau Extra Special	(88.7)	6.9	0.05	0.3	Tr	2.4	59	244
215	**Mild**, draught	Mixed sample from different brewers	(95.0)	2.5	0.03	0.2	Tr	1.6	24	102
216	**Pale ale**, bottled	Mixed sample from different brewers	(93.9)	2.8	0.05	0.3	Tr	2.0	28	118
217	**Shandy**	10 cans, 4 brands	(94.0)	0.7	Tr	Tr	0	3.0	11	48
218	homemade	Recipe. Calculated using equal volumes of lager and lemonade	(94.4)	2.0	0.03	0.2	0	2.9	26	107
219	**Stout**, Guinness	10 samples; canned, bottled and draught	(90.2)	3.3	0.06	0.4	Tr	1.5	30	126
220	Mackeson	10 samples; canned and bottled	(92.0)	2.5	0.06	0.4	Tr	4.6	36	153
221	**Strong ale/barley wine**	Mixed sample from different brewers	(86.3)	5.7	0.11	0.7	Tr	6.1	66	275

[a] Includes 0.9g oligosaccharides
[b] Includes 0.3g oligosaccharides
[c] Includes 0.5g oligosaccharides

Carbohydrate fractions and fatty acids, g per 100ml
Cholesterol, mg per 100ml

No. 17-	Food	Starch g	Total sugars g	Individual sugars: Gluc g	Fruct g	Sucr g	Malt g	Lact g	Dietary fibre: Southgate method g	Englyst method g	Fatty acids: Satd g	Mono-unsatd g	Poly-unsatd g	Cholesterol mg
	Beers													
207	**Bitter**	0	2.2	0	0	0	2.2	0	Tr	Tr	Tr	Tr	Tr	0
208	best/premium	0	2.2	0.3	0	0	1.9	0	Tr	Tr	Tr	Tr	Tr	0
209	low alcohol	0	1.2	0.4	0.1	0	0.7	0	Tr	Tr	0	0	0	0
210	**Brown ale**, bottled	0	3.0	0.4	0.4	0.1	2.1	0	Tr	Tr	Tr	Tr	Tr	0
211	**Lager**	0	Tr	Tr	0	0	0	0	Tr	Tr	Tr	Tr	Tr	0
212	alcohol-free	0	1.2	0.6	0.4	Tr	0.2	0	Tr	Tr	Tr	Tr	Tr	0
213	low alcohol	0	1.0	0.5	0.2	0	0.3	0	Tr	Tr	Tr	Tr	Tr	0
214	premium	0	2.4	1.0	0	0	(1.4)	0	Tr	Tr	Tr	Tr	Tr	0
215	**Mild**, draught	0	1.6	Tr	Tr	0	1.6	0	Tr	Tr	Tr	Tr	Tr	0
216	**Pale ale**, bottled	0	2.0	0.7	Tr	0	1.3	0	Tr	Tr	Tr	Tr	Tr	0
217	**Shandy**	0	3.0	1.6	1.7	1.7	0	0	Tr	Tr	0	0	0	0
218	homemade	0	2.9	0.8	0.7	1.4	0.1	0	Tr	Tr	0	0	0	0
219	**Stout**, Guinness	0	1.5	Tr	Tr	0	(1.5)	Tr	N	N	Tr	Tr	Tr	0
220	Mackeson	0	4.6	0.4	0.3	0.4	(3.5)	0.9	N	N	Tr	Tr	Tr	0
221	**Strong ale/barley wine**	0	6.1	Tr	Tr	0	6.1	0	Tr	Tr	Tr	Tr	Tr	0

Inorganic constituents per 100ml

No. 17-	Food	mg										µg	
		Na	K	Ca	Mg	P	Fe	Cu	Zn	Cl	Mn	Se	I
	Beers												
207	**Bitter**	6	32	8	7	14	0.1	0.01	0.1	24	0.03	Tr	N
208	best/premium	8	46	9	8	16	Tr	0.03	0.1	36	0.01	Tr	N
209	low alcohol	7	34	10	6	8	Tr	Tr	Tr	1	0.01	Tr	N
210	**Brown ale**, bottled	16	33	7	6	11	0	0.07	0.3	37	Tr	Tr	N
211	**Lager**	7	39	5	7	19	Tr.	Tr	Tr	20	0.01	Tr	N
212	alcohol-free	2	44	3	7	19	Tr	Tr	Tr	Tr	0.01	Tr	N
213	low alcohol	12	56	8	12	10	Tr	Tr	Tr	1	0.01	Tr	N
214	premium	7	39	5	7	19	Tr	Tr	Tr	20	0.01	Tr	N
215	**Mild**, draught	11	33	10	8	12	0	0.05	Tr	34	Tr	Tr	N
216	**Pale ale**, bottled	10	49	9	10	15	0	0.04	Tr	31	Tr	Tr	N
217	**Shandy**	7	6	8	1	5	Tr	Tr	Tr	8	Tr	Tr	Tr
218	homemade	7	27	5	4	1	Tr	Tr	Tr	11	0.01	Tr	Tr
219	**Stout**, Guinness	6	48	4	8	26	0.2	Tr	Tr	17	0.01	Tr	N
220	Mackeson	18	37	4	6	14	Tr	Tr	Tr	33	Tr	Tr	N
221	**Strong ale/barley wine**	15	110	14	20	40	0	0.08	Tr	57	Tr	Tr	N

No. 17-	Food	Retinol µg	Carotene µg	Vitamin D µg	Vitamin E mg	Thiamin mg	Riboflavin mg	Niacin mg	Trypt/60 mg	Vitamin B_6 mg	Vitamin B_{12} µg	Folate µg	Pantothenate mg	Biotin µg	Vitamin C mg
	Beers														
207	**Bitter**	0	Tr	0	N	Tr	0.03	0.2	0.2	0.07	Tr	3	10.00	1	0
208	best/premium	0	Tr	0	N	Tr	0.04	0.8	0.2	0.09	Tr	8	0.07	1	0
209	low alcohol	0	Tr	0	N	Tr	0.02	0.7	0.2	0.05	Tr	3	0.09	Tr	0
210	**Brown ale**, bottled	0	Tr	0	N	Tr	0.02	0.3	0.1	0.01	Tr	4	0.10	1	0
211	**Lager**	0	Tr	0	N	Tr	0.04	0.7	0.3	0.06	Tr	12	0.03	1	0
212	alcohol-free	0	Tr	0	N	Tr	0.02	0.6	0.4	0.03	Tr	5	0.09	Tr	0
213	low alcohol	0	Tr	0	N	Tr	0.02	0.5	0.3	0.03	Tr	6	0.07	Tr	0
214	premium	0	Tr	0	N	Tr	0.04	0.7	0.3	0.06	Tr	12	0.03	1	0
215	**Mild**, draught	0	Tr	0	N	Tr	(0.03)	Tr	0.1	(0.02)	Tr	(5)	(0.10)	(1)	0
216	**Pale ale**, bottled	0	Tr	0	N	Tr	0.02	0.4	0.2	0.01	Tr	4	(0.10)	(1)	0
217	**Shandy**	0	Tr	0	N	Tr	Tr	0.1	Tr	0.01	Tr	1	0.02	Tr	0
218	homemade	0	Tr	0	N	Tr	0.02	0.3	0.2	0.03	Tr	6	0.02	Tr	0
219	**Stout**, Guinness	0	Tr	0	N	Tr	0.03	0.8	0.2	0.08	Tr	6	0.04	1	0
220	Mackeson	0	Tr	0	N	0.05	0.04	0.6	0.2	0.04	Tr	8	0.05	1	0
221	**Strong ale/barley wine**	0	Tr	0	N	Tr	0.06	0.8	0.4	0.04	Tr	9	N	N	0

No. 17-	Food	Description and main data sources	Water g	Alcohol g	Total Nitrogen g	Protein g	Fat g	Carbo-hydrate g	Energy value kcal	Energy value kJ
Ciders										
222	**Cider**, dry	3 samples of different brands	(92.5)	3.8	Tr	Tr	0	2.6	36	152
223	low alcohol	10 samples, 3 brands including Strongbow LA	94.9	0.6	Tr	Tr	0	3.6	17	74
224	sweet	3 samples of different brands	(91.2)	3.7	Tr	Tr	0	4.3	42	176
225	vintage	3 samples of the same brand	(80.6)	10.5	Tr	Tr	0	7.3	101	421
Wines										
226	**Champagne**	Mixed sample	(86.8)	9.9	0.04	0.3	0	1.4	76	315
227	**Mulled wine**, homemade	Recipe	(58.0)	14.4	0.02	0.1	0	25.2	196	823
228	**Red wine**	Mixed sample from different countries	(88.4)	9.6[a]	0.03	0.1	0	0.2	68	283
229	**Rosé wine**, medium	5 samples from different countries	(87.3)	8.7	0.01	0.1	0	2.5	71	294
230	**White wine**, dry	5 samples from different countries	(89.1)	9.1[b]	0.02	0.1	0	0.6	66	275
231	medium	Mixed sample from different countries	(86.3)	8.9[c]	0.02	0.1	0	3.0	74	308
232	sparkling	5 samples from different countries	(85.8)	7.6	0.04	0.3	0	5.1	74	307
233	sweet	Mixed sample from different countries	(80.6)	10.2	0.03	0.2	0	5.9	94	394
Fortified wines										
234	**Port**	2 samples	(71.1)	15.9	0.02	0.1	0	12.0	157	655
235	**Sherry**, dry	1 sample	(81.0)	15.7	0.03	0.2	0	1.4	116	481
236	medium	8 samples; including Spanish, British, Cyprus, own label	(78.8)	13.3	0.02	0.1	0	5.9	116	482
237	sweet	1 sample	(74.8)	15.6	0.05	0.3	0	6.9	136	568

[a] Typical range 8.7g to 10.7g (11.0ml to 13.5ml) alcohol per 100ml
[b] Typical range 7.1g to 10.3g (9.0ml to 13.0ml) alcohol per 100ml
[c] Typical range 7.9g to 9.0g (10.0ml to 11.4ml) alcohol per 100ml

Carbohydrate fractions and fatty acids, g per 100ml
Cholesterol, mg per 100ml

No. 17-	Food	Starch g	Total sugars g	Individual sugars: Gluc g	Fruct g	Sucr g	Malt g	Lact g	Dietary fibre: Southgate method g	Englyst method g	Fatty acids: Satd g	Mono-unsatd g	Poly-unsatd g	Cholest-erol mg
	Ciders													
222	**Cider**, dry	0	2.6	0.6	0.5	0.7	0.8	0	0	0	0	0	0	0
223	low alcohol	0	3.6	0.7	1.4	1.4	0.1	0	0	0	0	0	0	0
224	sweet	0	4.3	1.0	0.7	1.2	1.3	0	0	0	0	0	0	0
225	vintage	0	7.3	1.8	1.3	2.0	2.3	0	0	0	0	0	0	0
	Wines													
226	**Champagne**	0	1.4	0.6	0.8	Tr	0	0	0	0	0	0	0	0
227	**Mulled wine**, homemade	0	25.2	4.2	4.2	16.8	0	0	0	0	0	0	0	0
228	**Red wine**	0	0.2	Tr	Tr	Tr	0	0	0	0	0	0	0	0
229	**Rosé wine**, medium	0	2.5	0.8	1.7	0	0	0	0	0	0	0	0	0
230	**White wine**, dry	0	0.6	0.3	0.3	0	0	0	0	0	0	0	0	0
231	medium	0	3.0	1.2	1.4	N	0	0	0	0	0	0	0	0
232	sparkling	0	5.1	2.2	2.8	0.1	0	0	0	0	0	0	0	0
233	sweet	0	5.9	2.6	3.3	0.1	0	0	0	0	0	0	0	0
	Fortified wines													
234	**Port**	0	12.0	4.6	4.6	2.8	0	0	0	0	0	0	0	0
235	**Sherry**, dry	0	1.4	0.7	0.7	0	0	0	0	0	0	0	0	0
236	medium	0	5.9	3.0	2.9	0	0	0	0	0	0	0	0	0
237	sweet	0	6.9	3.6	3.5	0	0	0	0	0	0	0	0	0

Inorganic constituents per 100ml

No. 17-	Food	Na (mg)	K	Ca	Mg	P	Fe	Cu	Zn	Cl	Mn	Se (µg)	I
	Ciders												
222	**Cider**, dry	7	72	8	3	3	0.5	0.04	Tr	6	Tr	Tr	N
223	low alcohol	3	81	7	2	4	0.1	0.03	Tr	2	0.01	Tr	Tr
224	sweet	7	72	8	3	3	0.5	0.04	Tr	6	Tr	Tr	N
225	vintage	2	97	5	4	9	0.3	0.02	Tr	5	Tr	Tr	N
	Wines												
226	**Champagne**	4	57	3	6	7	0.5	0.01	Tr	7	Tr	Tr	N
227	**Mulled wine**, homemade	5	93	7	11	11	0.4	0.11	N	7	0.03	Tr	N
228	**Red wine**	7	110	7	11	13	0.9	0.06	0.1	11	0.10	Tr	N
229	**Rosé wine**, medium	4	75	12	7	6	1.0	0.02	Tr	7	0.10	Tr	N
230	**White wine**, dry	4	61	9	8	6	0.5	0.01	Tr	10	0.10	Tr	N
231	medium	11	81	12	8	8	0.8	Tr	Tr	3	0.10	Tr	N
232	sparkling	5	58	9	7	9	0.5	0.01	Tr	7	0.04	Tr	N
233	sweet	13	110	14	11	13	0.6	0.05	Tr	7	0.10	Tr	N
	Fortified wines												
234	**Port**	4	97	4	11	12	0.4	0.10	N	8	Tr	Tr	N
235	**Sherry**, dry	10	57	7	13	11	0.4	0.03	N	12	Tr	Tr	N
236	medium	27	55	8	5	24	0.4	0.04	Tr	10	0.01	Tr	N
237	sweet	13	110	7	11	10	0.4	0.11	N	14	Tr	Tr	N

No.	Food	Retinol	Carotene	Vitamin D	Vitamin E	Thiamin	Riboflavin	Niacin	Trypt/60	Vitamin B_6	Vitamin B_{12}	Folate	Pantothenate	Biotin	Vitamin C
17-		µg	µg	µg	mg	mg	mg	mg	mg	mg	µg	µg	mg	µg	mg
	Ciders														
222	**Cider**, dry	0	Tr	0	N	Tr	Tr	0	Tr	0.01	Tr	N	0.04	1	0
223	low alcohol	0	Tr	0	N	Tr	Tr	0.1	Tr	Tr	Tr	2	0.07	Tr	0
224	sweet	0	Tr	0	N	Tr	Tr	0	Tr	0.01	Tr	N	0.03	1	0
225	vintage	0	Tr	0	N	Tr	Tr	0	Tr	(0.01)	Tr	N	(0.03)	(1)	0
	Wines														
226	**Champagne**	0	Tr	0	N	Tr	0.01	0.1	Tr	0.02	Tr	Tr	0.03	N	0
227	**Mulled wine**, homemade	0	(2)	0	Tr	Tr	Tr	Tr	Tr	0.01	Tr	Tr	N	N	Tr
228	**Red wine**	0	Tr	0	N	Tr	0.02	0.1	Tr	0.03	Tr	1	0.04	2	0
229	**Rosé wine**, medium	0	Tr	0	N	Tr	0.01	0.1	Tr	0.02	Tr	Tr	0.04	N	0
230	**White wine** dry	0	Tr	0	N	Tr	0.01	0.1	Tr	0.02	Tr	Tr	0.03	N	0
231	medium	0	Tr	0	N	Tr	Tr	0.1	Tr	0.01	Tr	Tr	0.06	1	0
232	sparkling	0	Tr	0	N	Tr	0.01	0.1	Tr	0.02	Tr	Tr	0.04	1	0
233	sweet	0	Tr	0	N	Tr	0.01	0.1	Tr	0.01	Tr	Tr	0.03	N	0
	Fortified wines														
234	**Port**	0	Tr	0	0	Tr	0.01	0.1	Tr	0.01	Tr	Tr	N	N	0
235	**Sherry**, dry	0	Tr	0	0	Tr	0.01	0.1	Tr	0.01	Tr	Tr	N	N	0
236	medium	0	Tr	0	0	Tr	0.01	0.1	Tr	0.02	Tr	Tr	0.02	1	0
237	sweet	0	Tr	0	0	Tr	0.01	0.1	Tr	0.01	Tr	Tr	N	N	0

No. 17-	Food	Description and main data sources	Water g	Alcohol g	Total Nitrogen g	Protein g	Fat g	Carbo-hydrate g	Energy value kcal	Energy value kJ
	***Fortified wines** continued*									
238	**Tonic wine**	10 samples; Sanatogen	(73.7)	11.5	Tr	Tr	0	12.4	127	532
239	**Vermouth**, dry	5 samples of different brands	(82.1)	13.9	0.01	0.1	0	3.0	109	453
240	sweet	5 samples of different brands	(70.6)	13.0	Tr	Tr	0	15.9	151	631
	Liqueurs									
241	**Advocaat**	4 samples of different brands	(47.6)	12.8	0.75	4.7	6.3	28.4	260	1091
242	**Cream liqueurs**	2 samples of Baileys Original Irish Cream	(44.4)	13.5	Tr	Tr	16.1	22.8	325	1352
243	**Egg nog**	Recipe	(78.6)	3.6	0.60	3.8	4.1	9.8	114	477
244	**Liqueurs**, high strength	5 samples including Pernod, Drambuie, Cointreau, Grand Marnier, Southern Comfort	(28.0)	31.8	Tr	Tr	0	24.4	314	1313
245	low-medium strength	10 samples including Cherry Brandy, Tia Maria and Crème de Menthe	(47.4)	19.8[a]	Tr	Tr	0	32.8	262	1099
	Spirits									
246	**Spirits**, 37.5% volume	Mean of brandy, gin, rum, vodka and whisky	(70.4)	29.6	Tr	Tr	0	Tr	207	858
247	40% volume	Mean of brandy, gin, rum, vodka and whisky	(68.3)	31.7	Tr	Tr	0	Tr	222	919

[a] The alcohol content of Tia Maria is 20.9g, Pimms and Campari 19.8g, Malibu 19.0g, Monterez 13.8g per 100ml

Carbohydrate fractions and fatty acids, g per 100ml
Cholesterol, mg per 100ml

No.	Food	Starch	Total sugars	Individual sugars					Dietary fibre		Fatty acids			Cholest-erol
17-				Gluc	Fruct	Sucr	Malt	Lact	Southgate method	Englyst method	Satd	Mono-unsatd	Poly-unsatd	
		g	g	g	g	g	g	g	g	g	g	g	g	mg
***Fortified wines** continued*														
238	**Tonic wine**	0	12.4	6.5	5.9	0	0	0	0	0	0	0	0	0
239	**Vermouth**, dry	0	3.0	1.1	1.2	0.7	0	0	0	0	0	0	0	0
240	sweet	0	15.9	6.1	6.1	3.7	0	0	0	0	0	0	0	0
Liqueurs														
241	**Advocaat**	0	28.4	1.7	0	26.7	N	0	0	0	1.9	3.0	0.9	240
242	**Cream liqueurs**	Tr	22.8	0	0	22.0	0	0.8	0	0	N	N	N	N
243	**Egg nog**	0	9.8	0	0	6.3	0	3.5	0	0	2.1	1.4	0.2	56
244	**Liqueurs**, high strength	0	24.4	2.6	2.3	17.1	2.4	0	0	0	0	0	0	0
245	low-medium strength	0	32.8	6.3	6.1	20.4	0	0	0	0	0	0	0	0
Spirits														
246	**Spirits**, 37.5% volume	0	Tr	0	0	Tr	0	0	0	0	0	0	0	0
247	40% volume	0	Tr	0	0	Tr	0	0	0	0	0	0	0	0

No.	Food	mg										µg	
17-		Na	K	Ca	Mg	P	Fe	Cu	Zn	Cl	Mn	Se	I
	***Fortified wines** continued*												
238	**Tonic wine**	140	30	7	4	160	0.6	0.01	0.1	Tr	0.05	Tr	N
239	**Vermouth**, dry	11	34	7	6	6	0.3	0.03	Tr	7	Tr	Tr	N
240	sweet	28	30	6	4	6	0.4	0.04	Tr	16	Tr	Tr	N
	Liqueurs												
241	**Advocaat**	N	N	N	N	N	N	N	N	N	N	N	N
242	**Cream liqueurs**	89	19	18	2	38	0.1	Tr	0.2	25	Tr	Tr	N
243	**Egg nog**	57	120	91	1	91	0.3	0.02	0.5	92	Tr	2	17
244	**Liqueurs**, high strength	6	3	Tr	Tr	Tr	Tr	Tr	Tr	4	Tr	Tr	Tr
245	low-medium strength	12	34	5	2	7	0.1	0.02	Tr	20	0.02	N	N
	Spirits												
246	**Spirits**, 37.5% volume	Tr	Tr	Tr	Tr	Tr	Tr	Tr	Tr	Tr	Tr	Tr	Tr
247	40% volume	Tr	Tr	Tr	Tr	Tr	Tr	Tr	Tr	Tr	Tr	Tr	Tr

No. *17-*	Food	Retinol µg	Carotene µg	Vitamin D µg	Vitamin E mg	Thiamin mg	Riboflavin mg	Niacin mg	Trypt/60 mg	Vitamin B_6 mg	Vitamin B_{12} µg	Folate µg	Pantothenate mg	Biotin µg	Vitamin C mg
	***Fortified wines** continued*														
238	**Tonic wine**	0	0	0	N	Tr	Tr	0.5	Tr	0.01	Tr	Tr	0.03	1	0
239	**Vermouth**, dry	0	Tr	0	0	Tr	Tr	0	Tr	0.01	Tr	Tr	N	N	0
240	sweet	0	Tr	0	0	Tr	Tr	0	Tr	Tr	Tr	Tr	N	N	0
	Liqueurs														
241	**Advocaat**	N	N	Tr	N	N	N	N	1.4	N	N	N	N	N	0
242	**Cream liqueurs**	190	91	Tr	0.57	N	N	N	N	N	0	Tr	N	N	0
243	**Egg nog**	60	15	0.2	0.20	0.03	0.18	0.1	1.0	0.06	1	10	0.47	4	1
244	**Liqueurs**, high strength	0	Tr	0	0	Tr	Tr	Tr	Tr	Tr	Tr	Tr	Tr	Tr	0
245	low-medium strength	0	Tr	0	0	Tr	Tr	Tr	Tr	Tr	Tr	Tr	Tr	Tr	0
	Spirits														
246	**Spirits**, 37.5% volume	0	0	0	0	0	0	0	0	0	0	0	0	0	0
247	40% volume	0	0	0	0	0	0	0	0	0	0	0	0	0	0

SOUPS, SAUCES AND PICKLES

This section includes data on a variety of homemade soups, commercial canned and packet soups and a wide variety of sauces, dressings, chutneys and pickles. Most of the values in the fifth edition of *The Composition of Foods* have been updated and many new products have been included, but milk based sauces have not been included here as they were covered in the *Milk Products and Eggs* supplement.

Both concentrated and dried soups have been made up according to the manufacturers instructions but using distilled water. Since tap water can vary in composition by both area and source of supply, users may wish to contact their local water board for the composition of tap water in their specific area. Hard waters may contain as much as 160mg calcium and 50mg magnesium per litre.

Where the acetic acid content of products was known, its contribution has been included in the energy calculations.

The fatty acid profile of margarine used in the recipe calculations is an average of hard, soft and polyunsaturated varieties.

Losses of labile vitamins assigned to homemade soups were estimated from the following figures:

Typical percentage losses of vitamins in homemade soups

Thiamin	35
Riboflavin	20
Niacin	30
Vitamin B_6	40
Folate	40
Vitamin C	45

Composition of food per 100g

No. 17-	Food	Description and main data sources	Water g	Total Nitrogen g	Protein g	Fat g	Carbo-hydrate g	Energy value kcal	Energy value kJ
	Soups								
248	**Bouillabaisse**	Recipe	(77.9)	1.67	10.4	9.7	1.6	135	560
249	**Carrot and orange soup**	Recipe	94.3	0.07	0.4	0.5	3.7	20	85
250	**Chicken soup**, cream of, canned	10 cans, 3 brands	87.9	0.27	1.7	3.8	4.5	58	242
251	condensed	7 cans of the same brand	82.2	0.41	2.6	5.8	6.0	75	311
252	-, *as served*	Diluted with an equal volume of water	91.1	0.20	1.3	2.9	3.0	43	177
253	**Chicken noodle soup**, dried	10 packets, 4 brands	5.0	2.74	17.1	4.8	56.0[a]	322	1364
254	*as served*	Calculated from 35g soup powder to 570ml water	94.5	0.16	1.0	0.3	3.2[b]	19	79
255	**Consommé**	6 samples, 6 brands, including beef, game and chicken	95.2	0.46	2.9	Tr	0.1	12	51
256	**French onion soup**	Recipe	89.7	0.03	0.2	2.1	5.7	40	269
257	**Gazpacho**	Recipe	(93.0)	0.13	0.8	3.6	2.6	45	188
258	**Goulash soup**	Recipe	75.7	1.10	6.9	7.2	6.1	115	482
259	**Instant soup powder**	10 packets, 3 brands; assorted flavours	4.1	1.04	6.5	14.0	64.4[c]	393	1659
260	*as served*	Calculated from 37g powder to 190ml water	84.4	0.17	1.1	2.3	10.5[d]	64	270
261	calorie controlled	10 packets, 4 brands; assorted flavours	4.5	2.00	12.5	9.4	51.5[e]	328	1384
262	*as served*	Calculated from 37g powder to 190ml water	84.4	0.33	2.0	1.5	8.4[f]	53	226
263	**Lentil soup**, canned	10 cans, 4 brands	88.2	0.50	3.1	0.2	6.5	39	164
264	**Lentil soup**	Recipe	69.8	0.90	5.6	4.7	14.0[g]	117	493
265	**Low calorie soup**, canned	7 cans, 3 brands; tomato, vegetable and minestrone varieties	93.3	0.14	0.9	0.2	4.0	20	87

[a] Includes 5.8g maltodextrins
[b] Includes 0.3g maltodextrins
[c] Includes 18.7g maltodextrins
[d] Includes 3.0g maltodextrins
[e] Includes 2.7g maltodextrins
[f] Includes 0.2g maltodextrins
[g] Includes oligosaccharides

Carbohydrate fractions and fatty acids, g per 100g food
Cholesterol, mg per 100g food

No. 17-	Food	Starch g	Total sugars g	Individual sugars Gluc g	Fruct g	Sucr g	Malt g	Lact g	Dietary fibre Southgate method g	Englyst method g	Fatty acids Satd g	Mono-unsatd g	Poly-unsatd g	Cholesterol mg
	Soups													
248	**Bouillabaisse**	0.1	1.5	0.6	0.5	0.4	0	0	0.4	0.4	(1.3)	(6.4)	(0.9)	(32)
249	**Carrot and orange soup**	0.2	3.5	1.1	0.9	1.4	0	0.1	1.1	1.0	0.2	0.1	0.1	1
250	**Chicken soup**, cream of, canned	3.4	1.1	Tr	0.1	0.6	0	0.4	Tr	Tr	(0.6)	(2.0)	(1.0)	97
251	condensed	4.6	1.4	Tr	0.2	0.4	0	0.8	Tr	Tr	0.8	3.0	1.4	(4)
252	-, *as served*	2.3	0.7	Tr	0.1	0.2	0	0.4	0	0	0.4	1.5	0.7	(2)
253	**Chicken noodle soup**, dried	45.8	4.4	0.4	0.5	33.0	Tr	0	N	4.3	N	N	N	N
254	*as served*	2.7	0.3	Tr	Tr	0.2	Tr	0	N	0.2	N	N	N	N
255	**Consommé**	0	0.1	Tr	0.1	Tr	0	0	0	0	Tr	Tr	Tr	Tr
256	**French onion soup**	Tr	5.7	1.2	3.5	0.9	0	0	1.3	1.0	0.2	0.7	1.0	0
257	**Gazpacho**	0.1	2.5	1.1	1.2	0.2	0	0	N	0.6	0.5	2.5	0.3	0
258	**Goulash soup**	4.7	1.9	0.8	0.7	0.4	0	Tr	1.0	0.9	1.7	2.9	2.1	18
259	**Instant soup powder**	34.1	11.3	2.1	2.2	7.0	Tr	Tr	N	N	N	N	N	6
260	*as served*	5.6	1.8	0.3	0.4	1.1	Tr	0	N	N	N	N	N	1
261	calorie controlled	30.3	5.0	0.8	0.9	3.3	Tr	0	N	N	N	N	N	Tr
262	*as served*	4.9	0.8	0.1	0.1	0.5	Tr	0	N	N	N	N	N	0
263	**Lentil soup**, canned	5.3	1.2	0.5	0.7	Tr	0	0	N	1.2	N	N	N	0
264	**Lentil soup**	11.3	1.9[a]	0.5	0.5	0.7	0	0	2.5	1.4	3.0	1.1	0.3	12
265	**Low calorie soup**, canned	2.0	2.0	1.0	1.0	Tr	0	0	N	N	Tr	Tr	Tr	0

[a] Includes 0.2g galactose per 100g

Inorganic constituents per 100g food

No. 17-	Food	mg										µg	
		Na	K	Ca	Mg	P	Fe	Cu	Zn	Cl	Mn	Se	I
	Soups												
248	**Bouillabaisse**	(54)	(210)	45	(14)	160	0.7	(0.05)	0.4	N	(0.05)	N	N
249	**Carrot and orange soup**	150	85	13	2	10	0.2	0.01	(0.1)	170	0.05	Tr	(1)
250	**Chicken soup**, cream of, canned	400	41	27	5	27	0.4	0.02	0.3	610	Tr	Tr	(2)
251	condensed	710	(62)	(41)	(7)	(41)	(0.5)	(0.03)	(0.5)	1070	Tr	Tr	(4)
252	-, *as served*	350	(31)	(20)	(4)	(20)	(0.3)	(0.02)	(0.3)	530	Tr	Tr	(2)
253	**Chicken noodle soup**, dried	5100	250	72	59	250	3.7	0.17	1.2	7630	0.74	N	N
254	*as served*	300	14	4	3	15	0.2	0.01	0.1	440	0.04	N	N
255	**Consommé**	530	N	N	N	N	N	N	N	610	N	N	N
256	**French onion soup**	79	54	3	2	4	0.1	0.01	Tr	120	Tr	Tr	Tr
257	**Gazpacho**	130	180	11	8	21	0.3	0.04	0.1	190	N	Tr	N
258	**Goulash soup**	380	280	12	14	62	0.9	0.09	1.2	590	0.09	1	(4)
259	**Instant soup powder**	3440	610	48	27	200	1.7	0.17	0.7	4770	0.25	N	N
260	*as served*	560	100	8	4	33	0.3	0.03	0.1	780	0.04	N	N
261	calorie controlled	5820	630	280	40	310	9.7	(0.16)	1.1	8170	0.42	N	N
262	*as served*	950	100	45	7	51	1.6	(0.03)	0.2	1330	0.07	N	N
263	**Lentil soup**, canned	450	97	11	9	40	0.8	0.08	0.3	670	0.11	N	N
264	**Lentil soup**	90	210	16	20	80	1.8	0.20	0.7	150	N	N	N
265	**Low calorie soup**, canned	370	130	13	7	17	0.3	0.01	0.1	580	0.05	N	N

No. 17-	Food	Retinol µg	Carotene µg	Vitamin D µg	Vitamin E mg	Thiamin mg	Riboflavin mg	Niacin mg	Trypt/60 mg	Vitamin B_6 mg	Vitamin B_{12} µg	Folate µg	Pantothenate mg	Biotin µg	Vitamin C mg
	Soups														
248	**Bouillabaisse**	Tr	(36)	Tr	N	N	N	N	2.0	N	N	N	N	N	3
249	**Carrot and orange soup**	5	2945	0	0.24	0.03	0.01	0.1	0.1	0.04	Tr	4	0.11	Tr	2
250	**Chicken soup**, cream of, canned	(39)	(16)	Tr	(0.55)	0.01	0.03	0.2	0.3	0.01	Tr	(1)	(0.04)	(0)	0
251	condensed	(96)	(39)	0	(0.93)	(0.02)	0.04	0.6	0.5	(0.01)	Tr	(1)	(0.06)	(0)	0
252	-, *as served*	(48)	(20)	0	(0.46)	(0.01)	0.02	0.3	0.2	(0.01)	Tr	Tr	(0.03)	(0)	0
253	**Chicken noodle soup**, dried	Tr	0	0	N	0.23	0.09	3.1	3.1	N	Tr	N	N	N	0
254	*as served*	Tr	0	0	N	0.01	0.01	0.2	0.2	N	Tr	N	N	N	0
255	**Consommé**	0	0	0	0	N	N	N	0.5	N	N	N	N	N	0
256	**French onion soup**	0	(13)	0	(0.16)	0.02	0.01	0	0.1	0.02	0	2	Tr	1	5
257	**Gazpacho**	0	N	0	(0.90)	0.04	0.01	0.5	0.1	0.08	0	10	(0.16)	(1)	11
258	**Goulash soup**	Tr	65	Tr	(0.42)	0.07	0.06	1.1	1.4	0.17	1	12	0.34	(1)	9
259	**Instant soup powder**	0	N	0	N	(0.05)	(0.02)	(0.4)	1.0	N	0	N	N	N	0
260	*as served*	0	N	0	N	(0.01)	Tr	(0.1)	0.2	N	0	N	N	N	0
261	calorie controlled	0	N	0	N	N	N	N	2.0	N	Tr	N	N	N	0
262	*as served*	0	N	0	N	N	N	N	0.3	N	0	N	N	N	0
263	**Lentil soup**, canned	0	N	0	N	Tr	0.02	3.2	0.6	0.01	0	N	N	N	0
264	**Lentil soup**	43	(55)	Tr	N	0.09	0.04	0.5	0.8	0.10	Tr	6	0.34	N	1
265	**Low calorie soup**, canned	0	N	0	N	0.35	0.14	2.0	0.1	0.20	0	(10)	N	N	Tr

No. 17-	Food	Description and main data sources	Water g	Total Nitrogen g	Protein g	Fat g	Carbo-hydrate g	Energy value kcal	Energy value kJ
	Soups *continued*								
266	**Minestrone soup**, canned	Manufacturer's data (Heinz)	(92.1)	0.22	1.4	0.8	5.1	32	135
267	**Minestrone soup**	Recipe	86.1	0.29	1.8	3.0	7.6	63	264
268	dried	10 packets, 4 brands	5.9	1.60	10.0	5.6	57.6[a]	306	1299
269	-, *as served*	Calculated from 45g soup powder to 570ml water	92.6	0.17	0.7	0.4	4.2[b]	22	94
270	**Mushroom soup**, cream of, canned	10 cans, 3 brands	90.4	0.20	1.1	3.0	3.9	46	192
271	**Mulligatawny soup**	Recipe	81.9	0.23	1.4	6.8	8.2	97	403
272	**Oxtail soup**, canned	10 cans, 3 brands	88.5	0.38	2.4	1.7	5.1	44	185
273	dried	10 packets, 5 brands	3.0	2.81	17.6	10.5	51.0	356	1504
274	-, *as served*	Calculated from 45g soup powder to 570ml water	92.5	0.22	1.4	0.8	3.9	27	116
275	**Pea and ham soup**	Recipe	82.5	0.64	4.0	(2.1)	9.2	70	295
276	**Potato and leek soup**	Recipe	87.9	0.24	1.5	2.6	6.2	52	220
277	**Scotch broth**	Recipe	81.5	1.34	8.3	3.4	5.0	82	346
278	**Tomato soup**, cream of, canned	10 cans, 3 brands	84.2	0.13	0.8	3.0	5.9	52	219
279	condensed	7 cans, 2 brands	70.6	0.27	1.7	6.8	14.6	123	514
280	-, *as served*	Diluted with an equal volume of water	85.3	0.14	0.9	3.4	7.3	62	258
281	dried	10 packets, 4 brands; including cream of tomato	4.6	0.77	4.8	14.3	65.1[c]	392	1652
282	-, *as served*	Calculated from 58g soup powder to 570ml water	91.2	0.07	0.4	1.3	6.0[d]	36	151

[a] Includes 1.5g maltodextrins
[b] Includes 0.1g maltodextrins
[c] Includes 5.1g maltodextrins
[d] Includes 0.5g maltodextrins

No.	Food		Total	Individual sugars					Dietary fibre		Fatty acids			Cholest-
17-		Starch	sugars	Gluc	Fruct	Sucr	Malt	Lact	Southgate method	Englyst method	Satd	Mono-unsatd	Poly-unsatd	erol
		g	g	g	g	g	g	g	g	g	g	g	g	mg
	***Soups** continued*													
266	**Minestrone soup**, canned	3.6	1.4	N	N	N	N	0	N	0.6	0.4	0.2	0.1	2
267	**Minestrone soup**	6.0	1.6	0.5	0.6	0.4	0.1	Tr	(1.2)	0.9	N	N	N	N
268	dried	36.4	19.7	3.0	3.8	12.9	Tr	0	5.9	N	N	N	N	Tr
269	-, *as served*	2.7	1.4	0.2	0.3	0.9	Tr	0	0.5	N	N	N	N	0
270	**Mushroom soup**, cream of, canned	3.1	0.8	Tr	0.1	0.3	0	0.4	N	0.1	0.5	1.6	0.9	1
271	**Mulligatawny soup**	4.7	3.1	0.6	0.9	0.5	Tr	0.1	0.8	0.9	3.7	1.9	0.4	17
272	**Oxtail soup**, canned	4.2	0.9	0.2	0.2	0.5	Tr	0	N	0.1	0.6	0.6	0.2	(7)
273	dried	41.8	9.2	1.0	0.9	5.5	1.5	Tr	3.4	N	N	N	N	N
274	-, *as served*	3.2	0.7	0.1	0.1	0.4	0.1	0	0.3	N	N	N	N	N
275	**Pea and ham soup**	7.8	1.4	0.4	0.3	0.7	0	Tr	2.1	1.4	(1.2)	(0.5)	(0.3)	6
276	**Potato and leek soup**	4.6	1.5	0.2	0.2	0.2	Tr	0.8	1.0	0.8	1.6	0.6	0.2	7
277	**Scotch broth**	2.9	2.1	0.8	0.7	0.6	Tr	Tr	1.5	(1.2)	1.5	1.2	0.3	28
278	**Tomato soup**, cream of, canned	3.3	2.6	0.8	0.6	1.2	0	Tr	N	0.7	0.5	1.6	0.8	1
279	condensed	3.4	11.2	2.4	1.8	6.2	0	0.8	N	1.0	1.0	2.6	3.0	(1)
280	-, *as served*	1.7	5.6	1.2	0.9	3.1	0	0.4	N	0.5	0.5	1.3	1.5	Tr
281	dried	23.0	37.0	4.6	5.5	26.9	Tr	Tr	3.0	N	6.0	3.0	0.3	2
282	-, *as served*	2.1	3.4	0.4	0.5	2.5	Tr	Tr	0.3	N	0.6	0.3	Tr	Tr

Inorganic constituents per 100g food

No.	Food	mg										µg	
17-		Na	K	Ca	Mg	P	Fe	Cu	Zn	Cl	Mn	Se	I
	Soups *continued*												
266	**Minestrone soup**, canned	380	120	22	9	27	0.3	0.05	0.2	590	0.10	0	1
267	**Minestrone soup**	140	120	14	7	28	0.5	0.05	0.2	150	0.11	N	N
268	dried	6400	670	140	60	290	2.6	0.20	1.3	8540	0.57	Tr	(74)
269	-, *as served*	470	49	11	4	21	0.2	0.02	0.1	630	0.04	Tr	(5)
270	**Mushroom soup**, cream of, canned	470	55	30	4	30	0.3	0.04	0.3	750	Tr	1	(3)
271	**Mulligatawny soup**	120	96	24	8	25	1.3	0.07	0.3	180	0.15	(1)	N
272	**Oxtail soup**, canned	440	93	40	6	37	1.0	0.04	0.4	660	Tr	Tr	(1)
273	dried	5250	700	140	44	260	4.3	0.25	2.4	7670	N	N	N
274	-, *as served*	400	54	11	3	20	0.3	0.02	0.2	590	N	N	(5)
275	**Pea and ham soup**	94	190	13	20	55	1.0	(0.04)	(0.7)	110	(0.23)	N	N
276	**Potato and leek soup**	83	160	29	7	35	0.4	0.02	0.2	150	0.07	(1)	(4)
277	**Scotch broth**	36	250	21	14	94	1.1	0.08	1.6	50	0.12	(1)	N
278	**Tomato soup**, cream of, canned	400	190	17	8	20	0.4	0.06	0.2	640	0.10	Tr	(3)
279	condensed	830	(360)	(32)	(15)	(38)	(0.7)	(0.11)	0.3	1320	0.10	Tr	(5)
280	-, *as served*	410	(180)	(16)	(8)	(19)	(0.3)	(0.06)	0.2	660	0.10	Tr	(3)
281	dried	3100	840	99	40	280	1.2	0.30	0.5	4990	0.27	N	N
282	-, *as served*	290	78	9	4	26	0.1	0.03	0.1	460	0.03	N	N

No. 17-	Food	Retinol µg	Carotene µg	Vitamin D µg	Vitamin E mg	Thiamin mg	Riboflavin mg	Niacin mg	Trypt/60 mg	Vitamin B_6 mg	Vitamin B_{12} µg	Folate µg	Pantothenate mg	Biotin µg	Vitamin C mg
	Soups *continued*														
266	**Minestrone soup**, canned	5	335	0	0.40	0.03	0.01	0.3	0.3	0.04	0	5	0.06	Tr	0
267	**Minestrone soup**	13	570	0.1	0.32	0.05	0.01	0.3	0.3	0.06	Tr	9	0.11	Tr	5
268	dried	0	N	0	N	0.21	0.15	3.1	1.9	N	0	N	N	N	0
269	-, *as served*	0	N	0	N	0.02	0.01	0.2	0.1	N	0	N	N	N	0
270	**Mushroom soup**, cream of, canned	(40)	(16)	0	(0.54)	Tr	0.05	0.3	0.2	0.01	Tr	(2)	(0.10)	(1)	0
271	**Mulligatawny soup**	51	(835)	Tr	(0.27)	0.05	0.02	0.4	0.3	(0.04)	Tr	4	(0.10)	Tr	2
272	**Oxtail soup**, canned	0	0	0	(0.20)	0.02	0.03	0.7	0.5	0.03	0	1	(0.05)	(0)	0
273	dried	0	0	0	N	10.40[a]	0.30	3.5	3.8	N	0	N	N	N	0
274	-, *as served*	0	0	0	N	0.80[a]	0.02	0.3	0.3	N	0	N	N	N	0
275	**Pea and ham soup**	15	740	Tr	(0.32)	0.10	0.03	0.5	0.6	0.05	Tr	N	0.35	N	2
276	**Potato and leek soup**	26	170	0	0.27	0.07	0.04	0.2	0.3	0.12	Tr	12	0.17	1	3
277	**Scotch broth**	Tr	755	Tr	0.22	0.09	0.09	1.7	1.8	0.11	1	10	0.34	1	5
278	**Tomato soup**, cream of, canned	(40)	210	0	(1.40)	0.03	0.02	0.5	0.1	0.06	Tr	12	(0.12)	(1)	Tr
279	condensed	0	(400)	0	(3.49)	(0.06)	0.05	1.0	0.2	(0.10)	0	(10)	(0.24)	(1)	Tr
280	-, *as served*	0	(200)	0	(1.75)	(0.03)	0.03	0.5	0.1	(0.05)	Tr	(5)	(0.12)	(1)	Tr
281	dried	Tr	N	0	N	0.02	0.02	0.2	0.1	N	0	N	N	N	Tr
282	-, *as served*	0	N	0	N	Tr	Tr	Tr	Tr	N	0	N	N	N	0

[a] Derived from the flavouring agent

No. 17-	Food	Description and main data sources	Water g	Total Nitrogen g	Protein g	Fat g	Carbo-hydrate g	Energy value kcal	Energy value kJ
	Soups *continued*								
283	**Vegetable soup**	Recipe	90.1	0.14	0.9	4.0	3.2	52	216
284	canned	10 cans, 3 brands	87.8	0.22	1.4	0.6	9.9	48	204
285	dried	10 packets, 5 brands	6.1	1.90	11.6	3.9	57.3[a]	296	1258
286	-, *as served*	Calculated from 45g soup powder to 570ml water	93.1	0.14	0.9	0.3	4.2[b]	22	92
287	**"Wholesoup"**, canned	Analysis and manufacturer's information (Heinz); assorted varieties	87.2	0.45	2.8	0.2	7.4	41	173
	Sauces								
288	**Apple sauce**, homemade	Recipe	77.6	0.03	0.2	0.1	16.7	64	274
289	**Barbecue sauce**	Ref. Marsh (1980)	(75.5)	0.16	1.0	0.1	*23.4*	*93*	*395*
290	homemade	Recipe	(54.1)	0.45	2.8	21.0	21.6	281	1170
291	**Black bean sauce**	6 samples, 3 brands	70.8	1.24	7.1	2.1	12.2	93	394
292	**Brown sauce**, hot	9 bottles, 3 brands	68.3	0.19	1.2	0.1	27.9	120[c]	510[c]
293	sweet	10 bottles, 4 brands	68.2	0.19	1.2	0.1	22.2	98[c]	418[c]
294	**Chilli sauce**	8 samples; assorted types	71.7	0.21	1.3	0.8	17.7	79	335
295	**Cook-in-sauces**, canned	9 samples, 3 brands; assorted flavours	87.4	0.18	1.1	0.8	8.3	43	181
296	**Cranberry sauce**	8 samples, 4 brands	55.3	0.03	0.2	0.1	39.9	151	646
297	**Curry paste**	20 samples, 3 brands	36.8	0.75	4.7	21.3	6.7	236	975
298	**Curry sauce**, canned	10 samples, 4 brands; assorted flavours	81.4	0.24	1.5	5.0	7.1	78	324
299	**Dips**, sour-cream based	7 samples, 4 brands; assorted flavours	54.1	0.46	2.9	37.0	4.0[d]	360	1482

[a] Includes 2.5g maltodextrins
[b] Includes 0.1g maltodextrins
[c] Includes 9 kcal, 39 kJ from acetic acid
[d] Includes 2.0g maltodextrins

Carbohydrate fractions and fatty acids, g per 100g food
Cholesterol, mg per 100g food

No.	Food	Starch	Total sugars	Individual sugars					Dietary fibre		Fatty acids			Cholest-erol
17-				Gluc	Fruct	Sucr	Malt	Lact	Southgate method	Englyst method	Satd	Mono-unsatd	Poly-unsatd	
		g	g	g	g	g	g	g	g	g	g	g	g	mg
	***Soups** continued*													
283	**Vegetable soup**	1.4	1.8	0.6	0.5	0.7	0	Tr	1.1	0.9	2.6	0.9	0.2	11
284	canned	4.8	5.1	1.4	1.6	2.1	0	0	N	1.5	N	N	N	N
285	dried	43.2	11.6	2.5	2.8	6.3	Tr	0	N	N	N	N	N	0
286	-, *as served*	3.2	0.9	0.2	0.2	0.5	Tr	0	N	N	N	N	N	0
287	**"Wholesoup"**, canned	6.2	1.2	0.5	0.7	Tr	0	0	N	1.8	Tr	Tr	Tr	1
	Sauces													
288	**Apple sauce**, homemade	0	16.7	1.4	4.0	11.4	0	0	1.5	1.1	0	0	0.1	0
289	**Barbecue sauce**	0.1	(23.1)	N	N	N	0	0	N	0.5	0	0	0.1	0
290	homemade	0.1	21.5	2.2	2.1	17.2	0	Tr	1.1	1.1	13.1	5.6	1.0	55
291	**Black bean sauce**	1.1	11.1	2.1	2.6	6.4	0	0	N	2.0	N	N	N	0
292	**Brown sauce**, hot	3.7	24.2	7.5	8.0	8.7	0	0	N	0.7	Tr	Tr	Tr	0
293	sweet	2.6	19.6	7.3	8.3	4.0	0	0	N	0.7	Tr	Tr	Tr	0
294	**Chilli sauce**	0.4	17.3	7.8	9.5	Tr	0	0	N	1.1	Tr	Tr	Tr	0
295	**Cook-in-sauces**, canned	3.3	5.0	1.1	1.3	2.6	0	Tr	N	N	0.1	0.4	0.2	Tr
296	**Cranberry sauce**	0.6	39.3	17.9	17.9	3.5	0	0	N	N	Tr	Tr	Tr	0
297	**Curry paste**	4.3	2.4	0.9	1.5	0	0	0	N	N	N	N	N	0
298	**Curry sauce**, canned	3.4	3.7	1.1	1.9	0.7	0	Tr	N	N	N	N	N	Tr
299	**Dips**, sour-cream based	Tr	2.0	0.9	0.7	0.4	Tr	Tr	N	N	N	N	N	(60)

Inorganic constituents per 100g food

No. 17-	Food	mg										µg	
		Na	K	Ca	Mg	P	Fe	Cu	Zn	Cl	Mn	Se	I
	***Soups** continued*												
283	**Vegetable soup**	130	110	13	4	20	0.3	0.02	0.1	130	0.07	(1)	N
284	canned	430	110	12	8	29	0.4	0.04	0.2	660	0.07	N	(16)
285	dried	5100	690	150	62	250	3.3	0.25	1.2	7160	0.57	N	N
286	-, *as served*	370	50	11	5	18	0.2	0.02	0.1	520	0.04	N	N
287	**"Wholesoup"**, canned	420	120	9	10	33	0.6	0.08	0.3	660	0.12	1	1
	Sauces												
288	**Apple sauce**, homemade	2	62	4	2	5	0.1	0.03	Tr	1	Tr	Tr	Tr
289	**Barbecue sauce**	1190	240	17	23	27	0.6	0.11	0.2	1830	0.10	Tr	1
290	homemade	360	290	70	29	44	2.3	0.17	(0.6)	620	0.30	(1)	N
291	**Black bean sauce**	2510	150	86	46	84	5.7	0.20	0.7	4230	0.50	N	N
292	**Brown sauce**, hot	1500	330	35	53	21	1.2	0.10	0.2	1660	0.34	N	N
293	sweet	1420	(330)	(35)	(53)	(21)	(1.2)	(0.10)	(0.2)	1620	(0.34)	N	N
294	**Chilli sauce**	2620	140	17	15	28	2.8	0.07	0.1	3700	0.17	Tr	2
295	**Cook-in-sauces**, canned	940	130	7	5	20	0.4	0.03	0.1	620	0.07	N	N
296	**Cranberry sauce**	Tr	39	5	2	6	0.1	0.10	Tr	(10)	0.07	Tr	N
297	**Curry paste**	1520	520	150	75	110	12.8	0.29	0.8	3740	1.12	N	N
298	**Curry sauce**, canned	980	180	30	18	31	1.1	0.05	0.2	760	0.20	N	N
299	**Dips**, sour-cream based	330	130	72	10	79	0.4	0.98	0.9	N	0.10	Tr	N

No. 17-	Food	Retinol µg	Carotene µg	Vitamin D µg	Vitamin E mg	Thiamin mg	Riboflavin mg	Niacin mg	Trypt/60 mg	Vitamin B_6 mg	Vitamin B_{12} µg	Folate µg	Pantothenate mg	Biotin µg	Vitamin C mg
	***Soups** continued*														
283	**Vegetable soup**	38	830	0	0.25	0.06	(0.01)	0.2	0.1	0.06	Tr	7	(0.10)	Tr	3
284	canned	0	18	0	N	0.09	0.02	2.5	0.2	0.01	0	10	N	N	Tr
285	dried	0	N	0	N	0.05	0.02	0.4	1.9	N	0	N	N	N	N
286	-, *as served*	0	N	0	N	Tr	Tr	Tr	0.2	N	0	N	N	N	Tr
287	**"Wholesoup"**, canned	0	500	0	0.15	0.10	0.02	0.6	0.5	0.05	0	2	0.11	Tr	0
	Sauces														
288	**Apple sauce**, homemade	0	12	0	0.18	0.02	0.01	0.1	0.1	0.03	0	1	0	1	8
289	**Barbecue sauce**	0	505	0	0.91	0.03	0.02	0.4	0.1	0.04	0	5	0.10	1	3
290	homemade	195	125	0.2	0.83	0.10	Tr	0.7	0.7	0.15	Tr	13	0.11	N	4
291	**Black bean sauce**	0	N	0	N	1.14	0.10	0.4	1.1	N	Tr	N	N	N	Tr
292	**Brown sauce**, hot	0	40	0	N	0.13	0.09	0.1	0.2	0.10	0	8	N	N	Tr
293	sweet	0	(40)	0	N	(0.13)	(0.09)	(0.1)	(0.2)	(0.10)	(0)	(8)	N	N	Tr
294	**Chilli sauce**	0	570	0	1.97	0.01	0.09	0.6	0.3	0.10	0	10	0.22	1	8
295	**Cook-in-sauces**, canned	Tr	N	0	N	Tr	0.01	0.1	0.1	0.03	0	1	N	N	Tr
296	**Cranberry sauce**	0	N	0	N	(0.02)	(0.02)	(0.1)	Tr	(0.01)	0	N	N	N	Tr
297	**Curry paste**	0	N	0	N	0.09	0.13	1.8	N	N	Tr	N	N	N	0
298	**Curry sauce**, canned	0	N	0	N	Tr	0.03	0.1	0.2	0.02	0	N	N	N	Tr
299	**Dips**, sour-cream based	N	N	N	N	N	N	N	N	N	Tr	N	N	N	N

Composition of food per 100g

No. 17-	Food	Description and main data sources	Water g	Total Nitrogen g	Protein g	Fat g	Carbo-hydrate g	Energy value kcal	Energy value kJ
	Sauces *continued*								
300	**Dressing**, blue cheese	7 samples, 2 brands	38.4	0.31	2.0	46.3	8.7	457	1886
301	'fat free'	9 samples of the same brand (Kraft); assorted flavours	79.5	0.13	0.8	1.2	14.0	67	282
302	French	8 samples, 6 brands	33.3	0.02	0.1	49.4	4.5	462	1902
303	-, homemade	Recipe	(27.2)	0.04	0.3	72.3	0.2	652	2683
304	low fat	Manufacturer's data (Heinz); assorted flavours	(85.8)	0.26	1.6	3.3	9.3	71	298
305	oil and lemon	Recipe	21.6	0.04	0.3	70.6	2.8	647	2661
306	thousand island	7 samples, 4 brands	47.6	0.18	1.1	30.2	12.5	323	1336
307	-, reduced calorie	8 samples, 3 brands	65.8	0.11	0.7	15.2	14.7	195	810
308	yogurt-based	6 samples, 4 brands; assorted types	57.4	0.35	2.2	27.5	7.0	292	1209
309	-, homemade	Recipe	80.4	0.82	5.2	2.8	4.2	62	261
310	**Gravy instant granules**	7 samples, 3 brands	4.0	0.70	4.4	32.5	40.6	462	1927
311	*made up*	Calculated from 23.5g granules to 300ml water	93.0	0.05	0.3	2.4	3.0	34	142
312	**Guacamole**	Recipe	79.3	0.23	1.4	12.7	2.2[a]	128	531
313	**Hollandaise sauce**, homemade	Recipe	12.2	0.77	4.8	76.2	Tr	707	2912
314	**Horseradish sauce**	8 samples, 5 brands; creamed and plain samples	64.0	0.40	2.5	8.4[b]	17.9[b]	153	640
315	**Hot pepper sauce**	7 samples, 3 brands (Tabasco, Encona, Calypso)	87.3	0.26	1.6	1.5	1.7	26	110
316	**Mayonnaise**	6 samples, 5 brands	18.0	0.18	1.1	75.6	1.7	691	2843
317	homemade	Recipe	9.1	0.34	2.1	86.7	0.1	789	3247
318	reduced calorie	12 samples, 8 brands	59.5	0.16	1.0	28.1	8.2	288	1188

[a] Includes mannoheptulose present in avocados

[b] Creamed varieties have an average of 13g fat and 21g carbohydrate. Plain varieties have an average of 5g fat and 11g carbohydrate

Carbohydrate fractions and fatty acids, g per 100g food
Cholesterol, mg per 100g food

No. 17-	Food	Starch g	Total sugars g	Individual sugars Gluc g	Fruct g	Sucr g	Malt g	Lact g	Dietary fibre Southgate method g	Englyst method g	Fatty acids Satd g	Mono-unsatd g	Poly-unsatd g	Cholesterol mg
	***Sauces** continued*													
300	**Dressing**, blue cheese	1.0	7.7	4.0	3.5	0.2	0	Tr	0	0	N	N	N	41
301	'fat free'	3.7	10.3	3.3	1.9	5.1	0	Tr	N	N	N	N	N	0
302	French	0	4.5	2.0	2.2	0.3	0	0	0	0	N	N	N	0
303	-, homemade	0	0.2	0.1	0.1	0	0	0	0	0	10.3	52.6	6.1	0
304	low fat	3.1	6.0	N	N	N	N	N	Tr	Tr	0.5	1.5	1.1	5
305	oil and lemon	0	2.8	0.1	0.2	2.5	0	0	Tr	Tr	7.4	25.0	34.1	0
306	thousand island	1.7	10.8	5.0	3.3	2.5	0	0	N	0.4	N	N	N	29
307	-, reduced calorie	0	14.7	5.0	5.1	4.6	0	0	N	N	N	N	N	N
308	yogurt-based	2.4	7.2	2.6	2.4	2.2	0	N	N	N	N	N	N	N
309	-, homemade	0	4.2	Tr	0.1	Tr	0	4.1	N	N	1.5	0.8	0.3	1
310	**Gravy instant granules**	39.3	1.3	0.6	0.5	0.2	0	0	Tr	Tr	N	N	N	N
311	*made up*	2.9	0.1	Tr	Tr	Tr	0	0	Tr	Tr	N	N	N	N
312	**Guacamole**	Tr	1.3	0.6	0.5	0.1	0	0	N	2.5	2.7	7.9	1.5	0
313	**Hollandaise sauce**, homemade	Tr	Tr	Tr	Tr	Tr	0	Tr	Tr	Tr	48.0	19.9	3.2	485
314	**Horseradish sauce**	3.0	15.0	4.0	3.6	7.4	0	0[a]	N	2.5	1.1	3.8	3.2	14
315	**Hot pepper sauce**	0.8	0.9	0.4	0.5	Tr	0	0	Tr	Tr	N	N	N	0
316	**Mayonnaise**	0.4	1.3	0.1	0.1	1.1	0	0	0	0	11.1	17.3	43.9	75
317	homemade	0	0.1	Tr	Tr	Tr	0	0	Tr	Tr	12.8	62.3	7.3	130
318	reduced calorie	3.6	4.6	1.1	1.0	2.5	0	0	0	0	N	N	N	22

[a] Creamed varieties contain lactose

No.	Food	mg										µg	
17-		Na	K	Ca	Mg	P	Fe	Cu	Zn	Cl	Mn	Se	I
	***Sauces** continued*												
300	**Dressing**, blue cheese	1110	52	58	7	61	0.6	0.02	0.4	1400	0.10	1	6
301	'fat free'	1170	N	N	N	N	N	N	N	N	N	N	N
302	French	460	N	N	N	N	N	N	N	N	N	N	N
303	-, homemade	940[a]	N	6	14	12	0.9	0.02	(0.1)	1450[a]	N	Tr	N
304	low fat	510	90	48	16	44	0.4	0.02	0.2	780	Tr	1	7
305	oil and lemon	920[a]	62	5	6	6	0.5	0.02	Tr	1410[a]	0.03	Tr	9
306	thousand island	900	130	24	9	34	0.3	0.05	0.2	1390	0.07	1	5
307	-, reduced calorie	600	N	N	N	N	N	N	N	1220	N	N	N
308	yogurt-based	650	120	58	7	60	0.1	Tr	0.3	840	0.02	N	N
309	-, homemade	760	280	180	24	160	0.7	0.01	0.6	1210	0.05	(2)	(56)
310	**Gravy instant granules**	6330	150	22	15	71	0.5	0.24	0.3	10000	0.40	N	N
311	*made up*	460	10	1	1	5	Tr	0.02	Tr	730	Tr	N	N
312	**Guacamole**	140	370	1	20	33	0.4	0.13	0.3	220	0.16	Tr	2
313	**Hollandaise sauce**, homemade	1000	63	55	9	150	2.3	0.08	1.1	1550	0.12	(5)	(68)
314	**Horseradish sauce**	910	220	43	18	42	0.6	0.05	0.4	1710	0.18	N	N
315	**Hot pepper sauce**	1520	220	25	26	32	2.4	0.10	0.2	1780	0.14	N	N
316	**Mayonnaise**	450	16	8	1	27	0.3	0.02	0.1	750	Tr	N	35
317	homemade	330	N	19	7	61	1.2	0.03	0.5	510	0.05	2	N
318	reduced calorie	(940)	N	N	N	N	N	N	N	(1450)	N	N	N

[a] If no salt is used, will contain only a trace of Na and Cl

No. 17-	Food	Retinol µg	Carotene µg	Vitamin D µg	Vitamin E mg	Thiamin mg	Riboflavin mg	Niacin mg	Trypt/60 mg	Vitamin B_6 mg	Vitamin B_{12} µg	Folate µg	Pantothenate mg	Biotin µg	Vitamin C mg
	***Sauces** continued*														
300	**Dressing**, blue cheese	46	27	0.2	5.91	0.01	0.04	0	0.7	0.01	0	5	0.12	1	0
301	'fat free'	0	Tr	0	N	Tr	Tr	Tr	Tr	Tr	0	Tr	Tr	Tr	Tr
302	French	0	0	0	N	0	0	0	0	0	0	0	0	0	0
303	-, homemade	0	N	0	3.68	0	0.01	0.1	0	Tr	0	Tr	Tr	Tr	0
304	low fat	9	22	0.7	0.16	0.01	0.03	Tr	0.4	0.01	0	1	0.07	Tr	Tr
305	oil and lemon	0	55	0	0	0.01	0.02	0.1	0	0.01	0	3	0.02	Tr	8
306	thousand island	14	170	0.1	8.10	0.01	0.02	0.1	0.2	0.02	0	4	0.10	1	Tr
307	-, reduced calorie	0	N	0	N	N	N	N	N	N	Tr	N	N	N	Tr
308	yogurt-based	N	N	N	N	N	N	N	N	N	N	N	N	N	Tr
309	-, homemade	25	(74)	0	0.15	0.06	0.26	0.3	1.2	0.09	0	19	0.45	2	4
310	**Gravy instant granules**	N	Tr	Tr	N	N	N	N	0.8	N	Tr	Tr	N	N	0
311	*made up*	N	0	0	N	N	N	N	0.1	N	0	0	N	N	0
312	**Guacamole**	0	190	0	(2.42)	0.09	0.12	1.0	0.2	0.28	0	13	0.79	3	11
313	**Hollandaise sauce**, homemade	820	(405)	1.9	2.48	0.06	0.17	0.1	1.4	0.06	2	17	0.98	13	0
314	**Horseradish sauce**	Tr	Tr	Tr	N	N	N	N	N	N	Tr	N	N	N	Tr
315	**Hot pepper sauce**	0	N	0	N	N	N	N	N	N	0	N	N	N	Tr
316	**Mayonnaise**	86	100	0.3	18.99	0.02	0.07	Tr	0.3	0.01	1	4	N	N	Tr
317	homemade	62	N	0.6	(4.60)	0.04	0.06	0	0.6	0.04	1	16	0.54	6	1
318	reduced calorie	Tr	57	Tr	8.33	N	N	N	N	N	0	N	N	N	0

No. 17-	Food	Description and main data sources	Water g	Total Nitrogen g	Protein g	Fat g	Carbo-hydrate g	Energy value kcal	Energy value kJ
	Sauces *continued*								
319	**Mint sauce**	8 samples, 4 brands	68.7	0.26	1.6	Tr	21.5	101[a]	432[a]
320	homemade	Recipe	(80.9)	0.13	0.8	0.1	18.2	73	309
321	**Oyster sauce**	10 samples; assorted types	63.2	0.62	3.9	0.1	16.9	80	340
322	**Pasta sauce**, white	Recipe	75.5	1.43	7.8	11.2	1.3	138	570
323	tomato based	9 samples, 4 brands; assorted types	83.9	0.32	2.0	1.5	6.9	47	200
324	**Raita**	Recipe	84.1	0.68	4.3	2.1	5.7	58	242
325	**Redcurrant jelly**	7 samples, 5 brands	30.8	0.05	0.3	Tr	63.8	240	1026
326	**Salad cream**	3 samples of different brands	47.2	0.23	1.5	31.0	16.7	348	1440
327	reduced calorie	Analysis and manufacturers' data	N	0.16	1.0	17.2	9.4	194	804
328	**Sandwich spread**	Analysis and manufacturers' information	60.9	0.19	1.2	9.8	23.5	186	778
329	**Sauce**, dry, casserole mix	8 samples, 5 brands; including liver and bacon, sausage, beef Bourguignon mixes	7.3	1.55	9.7	3.9	68.2	330	1400
330	-, *made up*	Calculated from 70g powder to 568ml water	89.8	0.18	1.1	0.4	7.3	35	150
331	dry mix	7 samples, 3 brands; including parsley, onion and bread	7.0	1.49	9.3	6.5	64.5	338	1431
332	-, *made up*	Calculated from 60g powder to 568ml milk	80.3	0.61	3.8	4.1	10.3	91	381
333	tomato base, homemade	Recipe	85.4	0.23	1.4	5.0	5.5[b]	71	297
334	**Soy sauce**	8 samples, 4 brands; light and dark varieties	68.6	0.48	3.0	Tr	8.2	43	182
335	**Sweet and sour sauce**, canned	10 samples, 4 brands	81.7	0.06	0.4	0.1	10.6	44[c]	188[c]
336	take-away	7 samples purchased from Chinese restaurants	65.3	0.03	0.2	3.4	32.8	157[d]	666[d]

[a] Includes 14 kcal, 61 kJ from acetic acid
[b] Includes 0.4g oligosaccharides
[c] Includes 2 kcal, 8 kJ from acetic acid
[d] Includes [illegible]

Carbohydrate fractions and fatty acids, g per 100g food
Cholesterol, mg per 100g food

No. *17-*	Food	Starch g	Total sugars g	Individual sugars: Gluc g	Fruct g	Sucr g	Malt g	Lact g	Dietary fibre: Southgate method g	Englyst method g	Fatty acids: Satd g	Mono-unsatd g	Poly-unsatd g	Cholesterol mg
	***Sauces** continued*													
319	**Mint sauce**	0	21.5	4.9	4.8	11.8	0	0	N	N	Tr	Tr	Tr	0
320	homemade	0	18.2	0.1	0.1	17.9	0	0	N	N	Tr	Tr	Tr	0
321	**Oyster sauce**	N	N	N	N	N	N	0	N	N	Tr	Tr	Tr	N
322	**Pasta sauce**, white	0.1	1.2	0.1	0.1	0.1	0	1.1	1.2	0.6	(6.8)	(3.2)	(0.5)	39
323	tomato based	1.2	5.7	2.9	2.8	Tr	0	0	N	N	0.2	0.3	0.8	0
324	**Raita**	Tr	5.7	0.2	0.2	0	0	3.2	0.2	0.2	1.1	0.6	0.1	7
325	**Redcurrant jelly**	0	63.8	21.4	21.4	21.0	0	0	N	N	0	0	0	0
326	**Salad cream**	Tr	16.7	1.9	1.9	12.9	0	0	N	N	3.9	6.2	19.4	43
327	reduced calorie	0.2	9.2	2.5	2.3	4.4	0	0	N	N	2.5	4.7	9.1	7
328	**Sandwich spread**	3.4	20.1	N	N	N	0	Tr	N	0.7	1.1	3.5	4.8	16
329	**Sauce**, dry, casserole mix	56.0	12.2	3.8	2.3	4.8	0	1.3	N	N	N	N	N	0
330	-, *made up*	6.1	1.2	0.4	0.2	0.5	0	0.1	N	N	N	N	N	0
331	dry mix	51.7	12.8	1.8	2.4	7.1	0.7	0.8	N	N	N	N	N	0
332	-, *made up*	4.8	5.5	0.2	0.2	0.7	Tr	4.4	N	N	N	N	N	(13)
333	tomato base, homemade	0.3	4.8	1.9	1.9	0.9	0	0	1.2	1.0	0.7	3.4	0.6	0
334	**Soy sauce**	0.9	7.3	2.2	0.9	4.2	0	0	0	0	0	0	0	0
335	**Sweet and sour sauce**, canned	3.3	7.3	2.5	2.9	1.9	0	0	N	N	0.1	0	0	0
336	take-away	5.3	27.5	9.5	9.7	8.3	0	0	N	N	N	N	N	0

No.	Food	mg										µg	
17-		Na	K	Ca	Mg	P	Fe	Cu	Zn	Cl	Mn	Se	I
	***Sauces** continued*												
319	**Mint sauce**	690	210	120	46	27	7.4	0.30	0.2	1120	0.86	Tr	Tr
320	homemade	680	60	39	3	17	1.7	0.03	0.1	1050	0.24	Tr	1
321	**Oyster sauce**	4160	94	25	50	N	2.4	0.48	1.4	N	N	N	N
322	**Pasta sauce**, white	300	250	140	14	180	0.6	0.44	1.2	480	0.07	(7)	N
323	tomato based	410	490	23	21	42	0.7	0.16	0.2	830	0.10	N	N
324	**Raita**	350	250	150	18	130	0.2	0.01	0.5	570	0.14	1	44
325	**Redcurrant jelly**	(10)	120	5	2	8	0.3	0.05	0.1	(40)	0.07	N	N
326	**Salad cream**	1040	40	18	9	48	0.5	0.02	0.3	1620	0.10	N	11
327	reduced calorie	N	N	N	N	N	N	N	N	N	N	N	N
328	**Sandwich spread**	880	130	30	18	36	0.5	0.03	0.2	1370	0.10	1	4
329	**Sauce**, dry, casserole mix	5400	800	92	62	210	3.9	0.26	1.0	7470	0.54	N	N
330	-, *made up*	590	88	10	7	23	0.4	0.03	0.1	820	0.06	N	N
331	dry mix	2710	470	600	45	130	3.6	0.25	0.9	4390	0.82	N	N
332	-, *made up*	300	170	160	14	95	0.4	0.02	0.4	500	0.08	N	N
333	tomato base, homemade	49	330	20	13	28	0.6	0.10	0.2	120	0.10	Tr	(3)
334	**Soy sauce**	7120	180	17	37	47	2.4	0.01	0.2	10640	0.18	N	N
335	**Sweet and sour sauce**, canned	390	93	10	6	10	0.5	0.02	0.1	460	0.18	N	N
336	take-away	150	16	6	2	4	6.0	Tr	Tr	240	0.04	N	N

No. 17-	Food	Retinol µg	Carotene µg	Vitamin D µg	Vitamin E mg	Thiamin mg	Riboflavin mg	Niacin mg	Trypt/60 mg	Vitamin B_6 mg	Vitamin B_{12} µg	Folate µg	Pantothenate mg	Biotin µg	Vitamin C mg
	***Sauces** continued*														
319	**Mint sauce**	0	Tr	0	Tr	Tr	Tr	Tr	0.3	Tr	0	Tr	Tr	Tr	Tr
320	homemade	0	125	0	0.85	0.02	0.06	0.2	(0.3)	0	0	19	0	0	5
321	**Oyster sauce**	N	N	N	N	N	0.04	N	0.8	N	5	N	N	N	0
322	**Pasta sauce**, white	135	64	0.1	0.30	0.10	0.28	2.2	1.6	0.12	Tr	13	1.00	7	0
323	tomato based	0	(100)	0	N	0.06	0.50	0.1	0.3	0.06	0	10	N	N	Tr
324	**Raita**	19	76	0	0.05	0.05	0.19	0.2	0.9	0.08	0	15	0.43	2	1
325	**Redcurrant jelly**	0	(25)	0	N	N	N	N	0.1	N	0	N	N	N	Tr
326	**Salad cream**	9	17	0.2	(10.60)	N	N	N	0.3	0.03	1	3	N	N	0
327	reduced calorie	N	N	N	N	N	N	N	0.2	N	N	N	N	N	0
328	**Sandwich spread**	9	550	0.1	6.00	0.02	0.02	0.1	0.3	0.03	0	5	0.10	1	4
329	**Sauce**, dry, casserole mix	0	N	0	N	N	N	N	N	N	Tr	N	N	N	0
330	-, *made up*	0	N	0	N	N	N	N	N	N	0	N	N	N	0
331	dry mix	N	N	N	N	N	N	N	N	N	Tr	N	N	N	0
332	-, *made up*	N	N	N	N	N	N	N	N	N	Tr	N	N	N	0
333	tomato base, homemade	0	320	0	1.73	0.06	0.02	0.8	0.2	0.13	0	7	0.21	2	7
334	**Soy sauce**	0	0	0	N	0.05	0.13	3.4	1.4	N	0	11	N	N	0
335	**Sweet and sour sauce**, canned	0	N	0	N	0.11	Tr	0	0.1	N	0	N	N	N	N
336	take-away	0	N	0	N	0.11	Tr	0	Tr	N	0	N	N	N	N

No. 17-	Food	Description and main data sources	Water g	Total Nitrogen g	Protein g	Fat g	Carbo-hydrate g	Energy value kcal	Energy value kJ
	***Sauces** continued*								
337	**Tartare sauce**	10 samples, 4 brands	53.5	0.21	1.3	24.6	17.9	299[a]	1241[a]
338	**Tomato ketchup**	10 samples, 5 brands	68.0	0.26	1.6	0.1	28.6	115	489
339	**Vinegar**	4 samples including malt, cider and wine vinegar	N	0.07	0.4	0	0.6	22[b]	89[b]
340	**Worcestershire sauce**	7 samples, 3 brands	75.3	0.22	1.4	0.1	15.5	65	276
	Chutney								
341	**Chutney**, apple, homemade	Recipe	43.5	0.14	0.9	0.2	52.2[c]	201	858
342	mango, oily	10 assorted samples	34.8	0.06	0.4	10.9	49.5	285	1202
343	-, sweet	10 samples, 5 brands	44.3	0.11	0.7	0.1	48.3	189[d]	806[d]
344	mixed fruit	10 samples, 6 brands; assorted fruits	52.4	0.10	0.6	Tr	39.7	155[d]	663[d]
345	tomato	9 samples, 5 brands	63.8	0.19	1.2	0.2	31.0	128[e]	542[e]
346	-, homemade	Recipe	54.0	0.19	1.2	0.4	40.9[f]	162	690
	Pickles								
347	**Piccalilli**	9 samples, 4 brands; mild, saucy and sweet varieties	79.1	0.16	1.0	0.5	17.6[g]	84[h]	360[h]
348	**Pickle**, chilli, oily	10 assorted samples; whole contents	50.1	0.53	3.3	24.5	9.9	271	1121
349	lime, oily	10 assorted samples; whole contents	59.6	0.30	1.9	15.5	8.3	178	739
350	mango, oily	10 assorted samples; whole contents	54.8	0.29	1.8	15.0	9.4	177	736
351	mixed vegetables	9 samples, 4 brands	91.8	0.05	0.3	0.4	3.4	21[e]	92[e]
352	sweet	9 samples, 4 brands	60.7	0.10	0.6	0.1	36.0	141[d]	604[d]

[a] Includes 5 kcal, 22 kJ from acetic acid
[b] Includes 18 kcal, 73 kJ from acetic acid
[c] Includes 1.1g oligosaccharides
[d] Includes 3 kcal, 14 kJ from acetic acid
[e] Includes 4 kcal, 18 kJ from acetic acid
[f] Includes 0.5g oligosaccharides
[g] Carbohydrate values range from 6g to 21g
[h] Includes 10 kcal, 43 kJ from acetic acid

Carbohydrate fractions and fatty acids, g per 100g food
Cholesterol, mg per 100g food

No.	Food			Individual sugars					Dietary fibre		Fatty acids			
17-		Starch	Total sugars	Gluc	Fruct	Sucr	Malt	Lact	Southgate method	Englyst method	Satd	Mono-unsatd	Poly-unsatd	Cholest-erol
		g	g	g	g	g	g	g	g	g	g	g	g	mg
	***Sauces** continued*													
337	**Tartare sauce**	1.7	16.2	6.5	6.3	3.4	0	0	Tr	Tr	N	N	N	49
338	**Tomato ketchup**	1.1	27.5	5.9	6.4	15.2	0	0	N	0.9	Tr	Tr	Tr	0
339	**Vinegar**	0	0.6	0.3	0.3	0	0	0	0	0	0	0	0	0
340	**Worcestershire sauce**	0.8	14.7	4.0	4.7	6.0	0	0	0	0	Tr	Tr	Tr	0
	Chutney													
341	**Chutney**, apple, homemade	0.2	51.1	19.5	20.1	11.5	0	0	1.8	1.2	Tr	Tr	Tr	0
342	mango, oily	0.4	49.1	N	N	N	0	0	1.4	0.9	N	N	N	0
343	-, sweet	2.6	45.7	20.6	19.5	5.6	0	0	N	N	Tr	Tr	Tr	0
344	mixed fruit	2.3	37.4	16.8	17.2	3.4	0	0	N	N	Tr	Tr	Tr	0
345	tomato	2.9	28.1	13.6	14.2	0.3	0	0	N	1.3	Tr	Tr	0.1	0
346	-, homemade	0.1	40.3	15.4	15.8	9.1	0	0	1.9	1.4	0.1	0.1	0.1	0
	Pickles													
347	**Piccalilli**	2.8	14.8	6.5	6.8	1.5	0	0	1.9	1.0	0.1	0.1	0.3	0
348	**Pickle**, chilli, oily	N	N	N	N	N	N	0	4.7	N	N	N	N	0
349	lime, oily	N	N	N	N	N	N	0	5.4	N	N	N	N	0
350	mango, oily	N	N	N	N	N	N	0	3.9	N	N	N	N	0
351	mixed vegetables	0.1	3.3	1.4	1.6	0.3	0	0	N	1.6	0.1	0.1	0.2	0
352	sweet	2.1	33.9	11.2	11.8	10.9	0	0	1.5	1.2	Tr	Tr	Tr	0

Inorganic constituents per 100g food

No.	Food	mg										µg	
17-		Na	K	Ca	Mg	P	Fe	Cu	Zn	Cl	Mn	Se	I
	Sauces *continued*												
337	**Tartare sauce**	800	42	15	17	36	0.5	0.03	0.3	1540	0	1	8
338	**Tomato ketchup**	1630	350	13	19	31	0.3	0.05	0.1	1800	0.10	N	N
339	**Vinegar**	5	34	3	4	10	(0.1)	(0.01)	(0.1)	47	(0.01)	(1)	N
340	**Worcestershire sauce**	1200	600	190	73	31	10.1	0.21	0.4	2090	0.98	(1)	(1)
	Chutney												
341	**Chutney**, apple, homemade	110	130	13	6	16	0.5	0.07	0.1	180	0.07	1	1
342	mango, oily	1090	57	23	27	10	2.3	0.10	0.1	1720	0.10	N	N
343	-, sweet	1300	42	9	19	8	1.1	Tr	Tr	1530	0.07	N	N
344	mixed fruit	800	(210)	(25)	(13)	(20)	(0.9)	(0.08)	Tr	1280	(0.15)	Tr	Tr
345	tomato	410	300	14	12	27	0.6	0.09	0.2	790	0.12	N	N
346	-, homemade	130	290	22	11	33	0.9	0.09	(0.2)	230	(0.13)	1	2
	Pickles												
347	**Piccalilli**	1340	40	16	6	17	0.6	0.03	0.1	1330	0.10	N	N
348	**Pickle**, chilli, oily	4050	250	79	130	N	8.6	0.30	0.7	N	N	N	N
349	lime, oily	530	240	120	80	N	5.8	0.20	0.3	N	N	N	N
350	mango, oily	3280	170	58	55	N	8.6	0.40	0.3	N	N	N	N
351	mixed vegetables	620	N	N	N	N	N	N	N	1130	N	N	N
352	sweet	1610	94	15	6	12	0.6	Tr	0.1	1750	0.15	N	N

No. 17-	Food	Retinol µg	Carotene µg	Vitamin D µg	Vitamin E mg	Thiamin mg	Riboflavin mg	Niacin mg	Trypt 60 mg	Vitamin B_6 mg	Vitamin B_{12} µg	Folate µg	Pantothenate mg	Biotin µg	Vitamin C mg
	***Sauces** continued*														
337	**Tartare sauce**	24	150	0.2	10.10	0.02	0.02	Tr	0.3	0.02	0	4	0.12	1	2
338	**Tomato ketchup**	0	230	0	N	1.00	0.09	2.1	0.2	0.03	0	1	N	N	2
339	**Vinegar**	0	0	0	0	0	0	0	0	0	0	0	0	0	0
340	**Worcestershire sauce**	0	8	0	N	Tr	(0.01)	0.4	0.2	N	0	(1)	N	N	0
	Chutney														
341	**Chutney**, apple, homemade	0	11	0	0.14	0.03	0.01	0.2	0.1	0.06	0	1	0.02	Tr	4
342	mango, oily	0	130	0	N	0.02	0.03	0.1	Tr	N	0	N	N	N	1
343	-, sweet	0	N	0	N	0	0.01	0.2	0.8	N	0	N	N	N	Tr
344	mixed fruit	0	(57)	0	N	Tr	(0)	(0.7)	0.1	Tr	0	Tr	Tr	Tr	Tr
345	tomato	0	N	0	N	0.05	0.15	0.1	0.2	0.02	0	N	N	N	Tr
346	-, homemade	0	390	0	0.88	0.10	0.01	0.9	0.2	0.16	0	17	0.19	2	6
	Pickles														
347	**Piccalilli**	0	N	0	N	Tr	0.02	0.1	0.2	0.01	0	N	N	Tr	Tr
348	**Pickle**, chilli, oily	0	57	0	N	0.09	0.09	1.4	0.5	N	0	N	N	N	1
349	lime, oily	0	82	0	N	0.10	0.06	0.5	0.3	N	0	N	N	N	1
350	mango, oily	0	200	0	N	0.06	0.07	0.3	0.6	N	0	N	N	N	3
351	mixed vegetables	0	N	0	N	N	N	N	0.1	N	0	N	N	N	N
352	sweet	0	250	0	N	0.03	0.01	0.1	0.1	0.01	0	Tr	N	Tr	Tr

No. 17-	Food	Description and main data sources	Water g	Total Nitrogen g	Protein g	Fat g	Carbo-hydrate g	Energy value kcal	Energy value kJ
	***Pickles** continued*								
353	**Relish**, corn/cucumber/onion	9 samples, 5 brands	67.0	0.16	1.0	0.3	29.2	119[a]	510[a]
354	burger/chilli/tomato	9 samples, 4 brands	68.4	0.19	1.2	0.1	27.6	114[b]	485[b]

[a] Includes 3 kcal, 14 kJ from acetic acid

[b] Includes 4 kcal, 18 kJ from acetic acid

Carbohydrate fractions and fatty acids, g per 100g food
Cholesterol, mg per 100g food

No. 17-	Food	Starch g	Total sugars g	Individual sugars: Gluc g	Fruct g	Sucr g	Malt g	Lact g	Dietary fibre: Southgate method g	Englyst method g	Fatty acids: Satd g	Mono-unsatd g	Poly-unsatd g	Cholesterol mg
	***Pickles** continued*													
353	**Relish**, corn/cucumber/onion	3.6	25.6	8.1	8.5	9.0	0	0	N	1.2	Tr	0.1	0.2	0
354	burger/chilli/tomato	2.5	25.1	6.5	6.8	11.8	0	0	N	1.3	Tr	Tr	Tr	0

Inorganic constituents per 100g food

No.	Food	mg										µg	
17-		Na	K	Ca	Mg	P	Fe	Cu	Zn	Cl	Mn	Se	I
	***Pickles** continued*												
353	**Relish**, corn/cucumber/onion	340	110	13	9	24	0.3	0.07	0.2	660	0.07	N	N
354	burger/chilli/tomato	480	290	13	12	26	0.3	0.07	0.1	980	0.10	N	N

No. 17-	Food	Retinol µg	Carotene µg	Vitamin D µg	Vitamin E mg	Thiamin mg	Riboflavin mg	Niacin mg	Trypt/60 mg	Vitamin B_6 mg	Vitamin B_{12} µg	Folate µg	Pantothenate mg	Biotin µg	Vitamin C mg
	***Pickles** continued*														
353	**Relish**, corn/cucumber/onion	0	N	0	N	N	N	N	0.2	N	0	N	N	N	N
354	burger/chilli/tomato	0	N	0	N	0.06	0.05	0.2	0.2	N	0	N	N	N	N

MISCELLANEOUS FOODS

This section includes a selection of miscellaneous food items and ingredients. Further data for selected industrial food ingredients and additives are included in the Appendix on page 170. Most of the values in the fifth edition of *The Composition of Foods* have been updated and many new products have been included. This section of the book now contains values for 21 foods compared with 11 in the fifth edition.

An entry for distilled water has been included in this section mainly for use in recipe calculations. Since there is considerable variation in the mineral composition of tap water both by area of the country and source of supply, the user may wish to contact their local water board for more specific information. Hard waters may contain as much as 160mg calcium and 50mg magnesium per litre.

Composition of food per 100g

No. 17-	Food	Description and main data sources	Water g	Total Nitrogen g	Protein g	Fat g	Carbo-hydrate g	Energy value kcal	Energy value kJ
	Miscellaneous foods								
355	**Baking powder**	6 samples of the same brand	6.3	0.91	5.2	0	37.8	163	693
356	**Bicarbonate of soda**	As sodium bicarbonate	0	0	0	0	0	0	0
357	**Carob powder**	2 samples of the same brand	4.8	0.78	4.9	0.1	37.0	159	679
358	**Cream of tartar**	As potassium tartrate	0	0	0	0	0	239[a]	1005[a]
359	**Garlic purée**	8 samples, 4 brands	38.8	0.56	3.5	33.6	16.9	380	1573
360	**Gelatine**	Literature sources and Ref. Lewis and English (1990)	13.0	15.20	84.4	0	0	338	1435
361	**Meat extract**	Mixed sample including Bovril and own brands	39.0	6.64	40.4	0.6	3.2	179	760
362	**Mustard powder**	2 brands	5.0	4.62	28.9	28.7	20.7	452	1884
363	*made up*	Made up with an equal volume of water	52.5	2.31	14.5	14.4	10.4	226	942
364	smooth	10 samples, 7 types including English and French	63.7	1.14	7.1	8.2	9.7	139	579
365	wholegrain	9 samples, 5 brands	65.0	1.31	8.2	10.2	4.2	140	584
366	**Quorn**, myco-protein	Analysis and manufacturer's information (Marlow Foods); chunks and pieces	75.0	1.90[b]	11.8[c]	3.5	2.0[d]	86	362
367	**Salt**	2 samples	Tr	0	0	0	0	0	0
368	**Stock cubes**, beef	10 samples, 6 brands inlcuding Bovril, Oxo and own brands	6.1	2.85[e]	16.8[f]	9.2	N	N	N
369	chicken	7 samples, 4 brands including Oxo	5.8	2.50[g]	15.4[f]	15.4	9.9	237	990
370	vegetable	8 samples, 4 brands including Oxo	5.7	2.16	13.5	17.3	11.6	253	1055
371	**Stuffing mix**, dried	10 samples, 4 brands; assorted flavours	5.9	1.58	9.9	5.2	67.2	338	1436
372	*made up*	Made up according to packet directions	76.4	0.45	2.8	1.5	19.3	97	412
373	sage and onion, homemade	Recipe	56.5	0.87	5.2	14.8	20.4	231	962

[a] From tartaric acid
[b] Additional non protein nitrogen from chitin is present in variable amounts
[c] N x 6.22
[d] Includes oligosaccharides
[e] Includes 0.17g purine nitrogen
[f] (Total N - purine N) x 6.25
[g] Purine nitrogen forms about 10% of total nitrogen

Carbohydrate fractions and fatty acids, g per 100g food
Cholesterol, mg per 100g food

No. 17-	Food	Starch g	Total sugars g	Individual sugars Gluc g	Fruct g	Sucr g	Malt g	Lact g	Dietary fibre Southgate method g	Englyst method g	Fatty acids Satd g	Mono-unsatd g	Poly-unsatd g	Cholest-erol mg
	Miscellaneous foods													
355	**Baking powder**	37.8	0	0	0	0	0	0	0	0	0	0	0	0
356	**Bicarbonate of soda**	0	0	0	0	0	0	0	0	0	0	0	0	0
357	**Carob powder**	5.7	31.3	4.3	3.7	23.3	0	0	N	N	Tr	Tr	Tr	0
358	**Cream of tartar**	0	0	0	0	0	0	0	0	0	0	0	0	0
359	**Garlic purée**	0	16.9	6.4	6.8	3.7	0	0	N	N	N	N	N	0
360	**Gelatine**	0	0	0	0	0	0	0	0	0	0	0	0	0
361	**Meat extract**	2.8	0.4	0.2	0.2	Tr	0	0	0	0	N	N	N	N
362	**Mustard powder**	N	N	N	N	N	0	0	N	N	1.5	19.8	5.4	0
363	*made up*	N	N	N	N	N	0	0	N	N	0.8	9.9	2.7	0
364	smooth	1.9	7.8	3.4	2.9	1.5	0	0	N	N	0.5	5.8	1.6	0
365	wholegrain	0.3	3.9	2.0	1.9	Tr	0	0	N	4.9	0.6	7.2	1.9	0
366	**Quorn**, myco-protein	Tr	1.1	0.1	0	0	0.1	0.9	N	4.8	N	N	N	1
367	**Salt**	0	0	0	0	0	0	0	0	0	0	0	0	0
368	**Stock cubes**, beef	N	N	N	N	N	0	0	0	0	N	N	N	Tr
369	chicken	7.9	2.0	Tr	0.3	1.6	0	0	0	0	N	N	N	Tr
370	vegetable	9.4	2.2	0.1	0.7	1.4	0	0	Tr	Tr	N	N	N	0
371	**Stuffing mix**, dried	62.8	4.4	0.8	1.6	2.0	0	0	N	4.7	2.4	1.6	0.1	5
372	*made up*	18.0	1.3	0.2	0.5	0.6	0	0	N	1.3	0.8	0.5	Tr	1
373	sage and onion, homemade	14.6	5.8	N	N	N	N	N	2.4	1.7	4.5	5.9	3.7	76

Inorganic constituents per 100g food

No. 17-	Food	mg										µg	
		Na	K	Ca	Mg	P	Fe	Cu	Zn	Cl	Mn	Se	I
	Miscellaneous foods												
355	**Baking powder**	11800[a]	49	1130[a]	9	8430[a]	Tr	Tr	2.8	29	Tr	Tr	Tr
356	**Bicarbonate of soda**	27380	0	0	0	0	0	0	0	0	0	0	0
357	**Carob powder**	(45)	1040	390	52	65	3.2	0.27	0.6	70	0.50	N	N
358	**Cream of tartar**	0	20780	0	0	0	0	0	0	0	0	0	0
359	**Garlic purée**	2740	300	26	17	74	1.1	0.07	0.4	4570	0.17	(2)	(3)
360	**Gelatine**	330	7	250	15	32	2.1	0.05	0.2	N	0.13	19	6
361	**Meat extract**	4370	970	37	65	400	8.1	0.26	1.5	6550	0.08	N	N
362	**Mustard powder**	5	940	330	260	180	9.5	0.20	(6.5)	62	1.70	N	N
363	*made up*	3	470	170	130	90	4.8	0.10	(3.3)	31	0.85	N	N
364	smooth	2950	200	70	82	190	2.9	0.19	1.0	3550	0.70	N	N
365	wholegrain	1620	220	120	93	200	2.8	0.21	1.2	2210	0.70	N	N
366	**Quorn**, myco-protein	310	120	28	31	220	0.4	0.81	7.5	N	2.10	N	N
367	**Salt**	39300	89	(10)	76	(1)	0.3	0.08	(0.1)	59900	Tr	N	44[b]
368	**Stock cubes**, beef	14560	490	40	32	240	1.2	0.70	0.8	21010	0.20	N	N
369	chicken	16300	400	120	47	200	4.9	0.10	1.2	8850	0.27	N	N
370	vegetable	16800	390	47	44	120	2.8	0.05	0.4	9550	0.26	N	N
371	**Stuffing mix**, dried	1460	240	960	41	130	5.1	0.17	0.8	2820	1.00	N	N
372	*made up*	420	69	280	12	37	1.5	0.05	0.2	810	0.30	N	N
373	sage and onion, homemade	420	150	58	13	77	1.0	0.11	0.6	650	0.27	11	15

[a] The sodium, calcium and phosphorus content will depend on the brand.
[b] Iodised salt contains 3100µg iodine per 100g. Sea salt contains 50µg iodine per 100g

No. 17-	Food	Retinol µg	Carotene µg	Vitamin D µg	Vitamin E mg	Thiamin mg	Riboflavin mg	Niacin mg	Trypt/60 mg	Vitamin B_6 mg	Vitamin B_{12} µg	Folate µg	Pantothenate mg	Biotin µg	Vitamin C mg
	Miscellaneous foods														
355	**Baking powder**	0	0	0	Tr	Tr	Tr	Tr	1.0	Tr	0	Tr	Tr	Tr	0
356	**Bicarbonate of soda**	0	0	0	0	0	0	0	0	0	0	0	0	0	0
357	**Carob powder**	0	N	0	N	Tr	0.13	0.6	N	0.04	0	N	N	N	0
358	**Cream of tartar**	0	0	0	0	0	0	0	0	0	0	0	0	0	0
359	**Garlic purée**	0	Tr	0	N	0	0.10	0.1	0.8	N	0	N	N	N	Tr
360	**Gelatine**	0	0	0	0	Tr	Tr	Tr	Tr	Tr	0	Tr	Tr	Tr	0
361	**Meat extract**	N	0	0	N	9.70	8.50	87.0	3.0	0.57	8	1050	N	N	0
362	**Mustard powder**	0	N	0	N	N	N	N	8.5	N	0	0	N	N	0
363	*made up*	0	N	0	N	N	N	N	4.3	N	0	0	N	N	0
364	smooth	0	N	0	N	N	N	N	2.1	N	0	0	N	N	0
365	wholegrain	0	N	0	N	N	N	N	2.4	N	0	0	N	N	0
366	**Quorn**, myco-protein	0	0	0	0	36.60	0.15	0.3	N	Tr	0	7	0.14	9	0
367	**Salt**	0	0	0	0	0	0	0	0	0	0	0	0	0	0
368	**Stock cubes** beef	N	N	0	N	N	N	N	N	N	N	N	N	N	0
369	chicken	N	N	0	N	N	N	N	N	N	N	N	N	N	0
370	vegetable	0	N	0	N	N	N	N	N	N	Tr	N	N	N	0
371	**Stuffing mix**, dried	Tr	Tr	Tr	N	1.42	0.90	1.8	1.8	N	Tr	N	N	N	0
372	*made up*	0	0	0	N	0.31	0.22	0.5	0.5	N	0	N	N	N	0
373	sage and onion, homemade	145	120	1.4	1.48	0.12	0.07	0.8	1.2	0.10	Tr	13	0.30	4	2

Composition of food per 100g

No. 17-	Food	Description and main data sources	Water g	Total Nitrogen g	Protein g	Fat g	Carbo-hydrate g	Energy value kcal	Energy value kJ
	Miscellaneous foods *continued*								
374	**Tomato purée**	8 samples, 4 brands	74.9	0.80	5.0	0.3	14.2	76	323
375	**Tomatoes**, sun dried	10 samples of different brands bottled in olive and sunflower oils	28.1	0.53	3.3	51.3	5.4	495	2041
376	**Vegetable purée**	6 samples, 3 brands	70.9	1.74	3.6	4.4	4.3	70	293
377	**Water**, distilled	Included for recipe calculation	100.0	0	0	0	0	0	0
378	**Yeast**, bakers, compressed	Literature sources	70.0	2.02[a]	11.4[b]	0.4	1.1	53	226
379	dried	Literature sources	5.0	6.32[a]	35.6[b]	1.5	3.5	169	717
380	**Yeast extract**	Mixed sample including Marmite and own brands	26.7	6.78[c]	40.7[b]	0.4	3.5	180	763

[a] Purine nitrogen forms about 10% of total nitrogen.
[b] (Total N - purine N) x 6.25
[c] Includes 0.27g purine nitrogen

No. 17-	Food	Starch g	Total sugars g	Individual sugars: Gluc g	Fruct g	Sucr g	Malt g	Lact g	Dietary fibre: Southgate method g	Englyst method g	Fatty acids: Satd g	Mono-unsatd g	Poly-unsatd g	Cholesterol mg
	Miscellaneous foods *continued*													
374	**Tomato purée**	0.1	14.1	6.5	7.6	Tr	0	0	N	2.8	Tr	0.1	0.1	0
375	**Tomatoes**, sun dried	2.5	2.9	1.1	1.8	0	0	0	N	N	6.7	15.1	27.2	0
376	**Vegetable purée**	0.2	4.1	0.4	3.4	0.3	0	0	N	N	N	N	N	0
377	**Water**, distilled	0	0	0	0	0	0	0	0	0	0	0	0	0
378	**Yeast**, baker's, compressed	1.1	Tr	Tr	Tr	Tr	0	0	6.2	N	N	N	N	0
379	dried	(3.5)	Tr	Tr	Tr	Tr	0	0	(19.7)	N	N	N	N	0
380	**Yeast extract**	1.9	1.6	Tr	1.5	0.2	0	0	0	0	N	N	N	0

Inorganic constituents per 100g food

No. 17-	Food	mg										µg	
		Na	K	Ca	Mg	P	Fe	Cu	Zn	Cl	Mn	Se	I
	Miscellaneous foods *continued*												
374	**Tomato purée**	240[a]	1200	35	26	94	1.4	2.90	0.5	550	0.24	N	N
375	**Tomatoes**, sun dried	1000	460	31	27	46	1.9	0.82	0.8	1870	0.20	Tr	48
376	**Vegetable purée**	1200	1120	39	33	70	1.0	0.40	0.4	620	0.25	Tr	Tr
377	**Water**, distilled	0	0	0	0	0	0	0	0	0	0	0	0
378	**Yeast**, bakers, compressed	16	610	25	59	390	5.0	1.60	3.2	20	N	N	N
379	dried	(50)	(2000)	80	230	(1290)	20.0	5.00	8.0	N	N	N	N
380	**Yeast extract**	4300	2100	70	160	950	2.9	0.20	2.7	6630	0.19	N	49

[a] The sodium content of unsalted tomato purée is approximately 20mg per 100g

No. 17-	Food	Retinol µg	Carotene µg	Vitamin D µg	Vitamin E mg	Thiamin mg	Riboflavin mg	Niacin mg	Trypt/60 mg	Vitamin B_6 mg	Vitamin B_{12} µg	Folate µg	Pantothenate mg	Biotin µg	Vitamin C mg
	Miscellaneous foods *continued*														
374	**Tomato purée**	0	650	0	5.37	0.40	0.19	4.0	0.7	0.11	0	22	1.00	6	10
375	**Tomatoes**, sun dried	0	400	0	23.98	N	N	N	N	N	0	N	N	N	Tr
376	**Vegetable purée**	0	4680	0	N	0	0.02	0.7	0.6	0.01	0	10	N	N	3
377	**Water**, distilled	0	0	0	0	0	0	0	0	0	0	0	0	0	0
378	**Yeast**, bakers, compressed	0	Tr	0	Tr	0.71	1.70	11.0	2.0	0.60	Tr	1250	3.50	60	Tr
379	dried	0	Tr	0	Tr	2.33[a]	4.00	8.5	7.0	2.00	Tr	4000	11.00	200	Tr
380	**Yeast extract**	0	0	0	N	4.10	11.90	64.0	9.0	1.60	1	1150	N	N	0

[a] Value for bakers yeast. Brewers yeast contains 15.6mg thiamin per 100g

BABY FOODS

This section has largely been based on information provided by the main manufacturers (HJ Heinz Company Ltd, Cow & Gate Nutricia Ltd, Milupa Ltd, Farley Health Products Ltd, Robinsons Ltd and The Boots Company Ltd). Values for infant formulas and infant rusks are covered in the *Milk Products and Eggs* and *Cereals and Cereal Products* supplements respectively.

The layout of page 2 is slightly different from that in the other sections in this book and gives starch, total sugars, fatty acids totals, fibre fractions and cholesterol. There are, however no values for individual sugars for the majority of products.

Typical values are provided in the main tables, but vitamin levels can vary depending on fortification procedures and formulation may change in the future. Users may wish to check with individual manufacturers for specific information.

Composition of food per 100g

No. 17-	Food	Description and main data sources	Water g	Total Nitrogen g	Protein g	Fat g	Carbo-hydrate g	Energy value kcal	Energy value kJ
	Canned/bottled savouries								
381	**Meat based meal**	8 varieties including beef and vegetable casserole, Lancashire hotpot and cauliflower with lamb	N	0.54	3.4	3.0	8.6	73	306
382	**Egg/cheese based meal**	5 varieties including caulifower cheese, vegetable and egg savouries	N	0.59	3.7	3.4	9.7	82	344
383	**Fish based meal**	4 varieties including fisherman's pie	N	0.54	3.4	3.1	9.3	76	321
384	**Pasta based meal**	5 varieties including spaghetti bolognese and spaghetti hoops and sausage	N	0.50	3.1	3.0	8.5	71	300
385	**Poultry based meal**	18 varieties including chicken and vegetable casserole, chicken supreme, chicken risotto and farmhouse vegetables and turkey	N	0.56	3.5	3.0	8.8	74	311
386	**Vegetable based meal**	18 varieties including vegetable casserole, vegetable risotto and lentil bake	N	0.40	2.5	2.3	9.8	67	284
	Canned/bottled sauces								
387	**Pour over sauce**, meat based	3 varieties (beef, beef and vegetable, and chicken)	(85.9)	0.41	2.6	2.1	10.0	67	282
388	vegetable based	2 varieties (creamy vegetable, and tomato)	(85.1)	0.26	1.6	1.6	11.4	64	269

Carbohydrate fractions and fatty acids, g per 100g food
Cholesterol, mg per 100g food

No. 17-	Food	Starch	Total sugars	Dietary fibre		Fatty acids			Cholesterol
				Southgate method	Englyst method	Satd	Mono-unsatd	Poly-unsatd	
		g	g	g	g	g	g	g	mg
	Canned/bottled savouries								
381	**Meat based meal**	6.8	1.8	N	1.1	1.2	1.4	0.4	8
382	**Egg/cheese based meal**	7.7	2.0	N	1.0	1.7	0.9	0.7	5
383	**Fish based meal**	7.3	2.0	N	0.6	1.2	1.2	0.6	26
384	**Pasta based meal**	7.0	1.5	N	0.7	1.4	1.2	0.4	7
385	**Poultry based meal**	6.8	2.0	N	1.0	0.7	1.5	0.8	11
386	**Vegetable based meal**	7.6	2.2	N	1.8	0.4	1.0	0.9	0[a]
	Canned/bottled sauces								
387	**Pour over sauce**, meat based	6.8	1.5	N	0.2	0.5	1.0	0.4	5
388	vegetable based	6.1	5.3	N	0.5	0.3	0.8	0.5	1

[a] Creamed vegetables contain approximately 4mg cholesterol per 100g

Inorganic constituents per 100g food

No. 17-	Food	mg										µg	
		Na	K	Ca	Mg	P	Fe	Cu	Zn	Cl	Mn	Se	I
	Canned/bottled savouries												
381	**Meat based meal**	20	220	20	10	50	1.8[a]	0.06	0.6	30	0.20	1	2
382	**Egg/cheese based meal**	60	200	90	12	80	2.3[a]	0.04	0.3	90	0.10	1	6
383	**Fish based meal**	20	240	40	7	57	1.5[a]	0.03	0.3	70	Tr	3	15
384	**Pasta based meal**	55	180	20	10	40	1.5[a]	0.08	0.5	75	0.10	1	2
385	**Poultry based meal**	25	180	20	10	50	1.5[a]	0.06	0.3	45	0.10	1	3
386	**Vegetable based meal**	25	210	30	15	45	1.7[a]	0.10	0.3	35	0.10	1	1
	Canned/bottled sauces												
387	**Pour over sauce**, meat based	23	97	20	6	30	2.3[a]	0.05	0.2	43	0	Tr	2
388	vegetable based	86	180	37	9	38	2.3[a]	0.05	0.2	150	0.10	1	4

[a] With no fortification, Fe levels are approximately 1mg per 100g. When fortified, levels of Fe are approximately 2.3mg per 100g

No. 17-	Food	Retinol µg	Carotene µg	Vitamin D µg	Vitamin E mg	Thiamin mg	Riboflavin mg	Niacin mg	Trypt/60 mg	Vitamin B_6 mg	Vitamin B_{12} µg	Folate µg	Pantothenate mg	Biotin µg	Vitamin C mg
	Canned/bottled savouries														
381	**Meat based meal**	Tr	1365	Tr	0.24	0.05	0.03	0.5	0.7	0.08	0	6	0.12	1	5
382	**Egg/cheese based meal**	24	640	0.1	0.35	0.05	0.04	0.2	0.7	0.09	Tr	11	0.14	1	27[a]
383	**Fish based meal**	22	1215	0.1	0.46	0.03	0.04	0.2	0.8	0.07	0	5	0.14	1	23[a]
384	**Pasta based meal**	4	710	0	0.38	0.03	0.02	0.5	0.6	0.05	0	5	0.10	Tr	4
385	**Poultry based meal**	2	1390	0.2	0.33	0.05	0.02	0.6	0.6	0.08	Tr	7	0.17	1	3
386	**Vegetable based meal**	Tr[b]	1360	Tr[b]	0.50	0.05	0.02	0.3	0.4	0.07	Tr	10	0.12	Tr	26[a]
	Canned/bottled sauces														
387	**Pour over sauce**, meat based	4	57	0	0.46	0.02	0.02	0.3	0.4	0.03	0	2	0.09	Tr	0
388	vegetable based	12	165	0.1	0.67	0.03	0.03	0.3	0.3	0.05	0	4	0.11	1	33

[a] Fortified values

[b] Creamed vegetables contain approximately 17µg retinol and 0.05µg vitamin D per 100g

No. 17-	Food	Description and main data sources	Water g	Total Nitrogen g	Protein g	Fat g	Carbo-hydrate g	Energy value kcal	Energy value kJ
	Canned/bottled desserts								
389	**Cereal and milk based desserts**	9 varieties including chocolate pudding, egg custard and rice pudding	N	0.47	3.0	2.6	13.2	85	358
390	**Fromage frais**, fruit	Assorted varieties including tropical fruit, mandarin and blackcurrant	N	0.19	1.2	1.9	10.5	61	259
391	**Fruit desserts**	21 varieties	N	0.08	0.5	0.2	15.4	62	262
392	**Fruit and milk based desserts**	8 varieties including fruit fool and fruit custard	N	0.28	1.8	1.9	14.2	78	328
393	**Yogurt**, fruit	7 varieties including apple, banana and strawberry	N	0.20	1.3	0.9	13.8	65	276
	Instant/granulated savouries								
394	**Meat based meal**	6 varieties including beef casserole, cottage pie and vegetables and lamb	N	2.62	16.4	8.0	66.7	388	1642
395	*reconstituted*	Calculated from 1g of product to 3ml water	N	0.66	4.1	2.0	16.7	97	411
396	**Egg/cheese based meal**	7 varieties including cauliflower cheese, cheese and spinach mornay and egg savoury	N	2.29	14.3	10.0	67.3	400	1690
397	*reconstituted*	Calculated from 1g of product to 3ml water	N	0.57	3.6	2.5	16.8	100	423
398	**Poultry based meal**	4 varieties including chicken casserole and chicken and country vegetables	N	2.74	17.1	8.2	65.6	388	1644
399	*reconstituted*	Calculated from 1g of product to 3ml water	N	0.69	4.3	2.1	16.4	97	411

Carbohydrate fractions and fatty acids, g per 100g food
Cholesterol, mg per 100g food

No. 17-	Food	Starch g	Total sugars g	Dietary fibre: Southgate method g	Dietary fibre: Englyst method g	Fatty acids: Satd g	Fatty acids: Mono-unsatd g	Fatty acids: Poly-unsatd g	Cholest-erol mg
	Canned/bottled desserts								
389	**Cereal and milk based desserts**	4.2	9.0	N	0.2	0.7	1.2	0.7	7
390	**Fromage frais**, fruit	5.2	5.3	N	0.2	1.1	0.6	0.1	6
391	**Fruit desserts**	2.0	13.4	N	1.0	Tr	0	0.1	0
392	**Fruit and milk based desserts**	4.2	10.0	N	0.4	0.5	0.9	0.5	14
393	**Yogurt**, fruit	5.1	8.7	N	0.3	0.1	0.4	0.2	1
	Instant/granulated savouries								
394	**Meat based meal**	N	10.4	N	N	N	N	N	N
395	*reconstituted*	N	2.6	N	N	N	N	N	N
396	**Egg/cheese based meal**	N	13.3	N	N	N	N	N	N
397	*reconstituted*	N	3.3	N	N	N	N	N	N
398	**Poultry based meal**	N	13.0	N	N	N	N	N	N
399	*reconstituted*	N	3.3	N	N	N	N	N	N

Inorganic constituents per 100g food

No. 17-	Food	mg										µg	
		Na	K	Ca	Mg	P	Fe	Cu	Zn	Cl	Mn	Se	I
	Canned/bottled desserts												
389	**Cereal and milk based desserts**	40	140	90	12	80	0.2	0.02	0.3	75	Tr	1	11
390	**Fromage frais**, fruit	8	80	16	6	22	0.2	0.02	0.1	20	0.10	Tr	1
391	**Fruit desserts**	5	170	20	8	20	0.3	0.03	0.1	8	0.10	Tr	1
392	**Fruit and milk based desserts**	20	150	50	7	50	0.2	0.02	0.2	40	0.10	1	6
393	**Yogurt**, fruit	19	150	39	11	41	0.3	0.02	0.2	40	0.10	Tr	12
	Instant/granulated savouries												
394	**Meat based meal**	300	650	400	70	250	6.0[a]	Tr	4.0	550	Tr	N	Tr
395	*reconstituted*	75	160	100	18	63	1.5[b]	0	1.0	140	0	Tr	0
396	**Egg/cheese based meal**	300	560	530	50	300	6.0[a]	Tr	4.0	270	Tr	N	Tr
397	*reconstituted*	75	140	130	13	75	1.5[b]	0	1.0	68	0	Tr	0
398	**Poultry based meal**	310	800	550	55	350	6.0[a]	Tr	4.0	590	Tr	N	Tr
399	*reconstituted*	78	200	140	14	88	1.5[b]	0	1.0	150	0	Tr	0

[a] With high fortification, levels of Fe are approximately 16mg per 100g food [b] With high fortification, levels of Fe are approximately 4mg per 100g food

No. 17-	Food	Retinol µg	Carotene µg	Vitamin D µg	Vitamin E mg	Thiamin mg	Riboflavin mg	Niacin mg	Trypt/60 mg	Vitamin B_6 mg	Vitamin B_{12} µg	Folate µg	Pantothenate mg	Biotin µg	Vitamin C mg
	Canned/bottled desserts														
389	**Cereal and milk based desserts**	14	0	0.1	0.58	0.02	0.08	0.1	0.7	0.03	0	3	0.15	1	0
390	**Fromage frais**, fruit	12	21[a]	0	0.05	0.02	0.03	0.1	0.3	0.03	0	3	0.03	Tr	29[b]
391	**Fruit desserts**	0	31	0	0.32	0.02	0.01	0.1	0.1	0.05	0	2	0.04	1	12[c]
392	**Fruit and milk based desserts**	17	12	0.1	0.48	0.02	0.04	0.1	0.4	0.03	0	3	0.11	1	27
393	**Yogurt**, fruit	2	15[d]	Tr	0.17	0.02	0.03	0.2	0.3	0.04	Tr	3	0.09	1	35[b]
	Instant/granulated savouries														
394	**Meat based meal**	N[e]	N	7.0[b]	3.70[b]	0.90[b]	0.50[b]	4.8[b]	3.3	0.60[b]	1	74[b]	2.60	7	45[f]
395	*reconstituted*	N[g]	N	1.8[b]	0.93[b]	0.23[b]	0.13[b]	1.2[b]	0.8	0.15[b]	Tr	19[b]	0.65	2	11
396	**Egg/cheese based meal**	N[h]	N	7.4[b]	3.70[b]	0.90[b]	0.50[b]	4.8[b]	2.7	0.60[b]	1	74[b]	2.60	7	44[f]
397	*reconstituted*	N[i]	N	1.9[b]	0.93[b]	0.23[b]	0.13[b]	1.2[b]	0.7	0.15[b]	Tr	19[b]	0.65	2	11
398	**Poultry based meal**	N[h]	N	7.4[b]	3.70[b]	0.90[b]	0.50[b]	4.8[b]	3.2	0.60[b]	1	74[b]	2.60	7	44[f]
399	*reconstituted*	N[j]	N	1.9[b]	0.93[b]	0.23[b]	0.13[b]	1.2[b]	0.8	0.15[b]	Tr	19[b]	0.65	2	11

[a] Levels ranged from 1µg to 107µg per 100g [b] Fortified values [c] With no fortification, vitamin C levels are approximately 5µg per 100g. When fortified, levels of vitamin C are approximately 30mg per 100g [d] Levels ranged from 1µg to 63µg per 100g [e] Vitamin A is 400µg per 100g food
[f] Levels ranged from 20mg to 60mg vitamin C per 100g [g] Vitamin A is 100µg per 100g food [h] Vitamin A is 390µg per 100g food
[i] Vitamin A is 98µg per 100g food

Composition of food per 100g

No. 17-	Food	Description and main data sources	Water g	Total Nitrogen g	Protein g	Fat g	Carbo-hydrate g	Energy value kcal	Energy value kJ
***Instant/granulated savouries** continued*									
400	**Vegetable based meal**	7 varieties including vegetable bake, vegetable casserole and vegetable hotpot	N	2.45	15.3	8.5	66.4	387	1637
401	*reconstituted*	Calculated from 1g of product to 3ml water	N	0.61	3.8	2.1	16.6	97	409
Instant/granulated sauces									
402	**Pour over sauce**, vegetable based	Assorted varieties including cheese, tomato, and white sauces	N	1.92	12.0	8.7	73.4	402	1700
403	*reconstituted*	Calculated from 1g product to 3ml water	N	0.48	3.0	2.2	18.4	100	425
Instant/granulated desserts									
404	**Cereal and milk based desserts**	7 varieties including rice pudding and semolina with honey	N	2.16	13.5	8.0	72.0	396	1678
405	*reconstituted*	Calculated from 1g of product to 2.5ml water	N	0.62	3.9	2.3	20.6	113	479
406	**Fruit desserts**	7 varieties including fruit cocktail and fruit salad	N	1.71	10.7	6.9	76.4	391	1660
407	*reconstituted*	Calculated from 1g of product to 2.5ml water	N	0.49	3.1	2.0	21.8	112	474
408	**Yogurt**, fruit	4 varieties including apricot, banana and strawberry	N	2.10	13.1	8.4	72.3	399	1690
409	*reconstituted*	Calculated from 1g of product to 2.5ml water	N	0.60	3.7	2.4	20.7	114	483

Carbohydrate fractions and fatty acids, g per 100g food
Cholesterol, mg per 100g food

No.	Food	Starch	Total sugars	Dietary fibre		Fatty acids			Cholest-erol
				Southgate method	Englyst method	Satd	Mono-unsatd	Poly-unsatd	
17-		g	g	g	g	g	g	g	mg
	Instant/granulated savouries *continued*								
400	**Vegetable based meal**	N	20.0	N	N	N	N	N	N
401	*reconstituted*	N	5.0	N	N	N	N	N	N
	Instant/granulated sauces								
402	**Pour over sauce**, vegetable based	N	N	N	N	N	N	N	N
403	*reconstituted*	N	N	N	N	N	N	N	N
	Instant/granulated desserts								
404	**Cereal and milk based desserts**	N	41.0	N	N	N	N	N	N
405	*reconstituted*	N	11.7	N	N	N	N	N	N
406	**Fruit desserts**	N	33.7	N	N	N	N	N	N
407	*reconstituted*	N	9.6	N	N	N	N	N	N
408	**Yogurt**, fruit	N	38.0	N	N	N	N	N	N
409	*reconstituted*	N	10.9	N	N	N	N	N	N

Inorganic constituents per 100g food

No. 17-	Food	Na (mg)	K (mg)	Ca (mg)	Mg (mg)	P (mg)	Fe (mg)	Cu (mg)	Zn (mg)	Cl (mg)	Mn (mg)	Se (µg)	I (µg)
	Instant/granulated savouries *continued*												
400	**Vegetable based meal**	220	750	540	70	390	6.0[a]	Tr	4.0	510	Tr	N	Tr
401	*reconstituted*	55	190	140	18	96	1.5[b]	0	1.0	130	0	Tr	0
	Instant/granulated sauces												
402	**Pour over sauce**, vegetable based	N	N	N	N	N	N	N	N	N	N	N	N
403	*reconstituted*	N	N	N	N	N	N	N	N	N	N	N	N
	Instant/granulated desserts												
404	**Cereal and milk based desserts**	120	450	500	30	250	6.0	Tr	4.0	200	Tr	Tr	Tr
405	*reconstituted*	33	130	140	9	71	1.7	0	1.1	57	0	0	0
406	**Fruit desserts**	110	510	540	60	330	5.9	Tr	4.0	250	Tr	Tr	Tr
407	*reconstituted*	31	150	150	17	93	1.7	0	1.1	71	0	0	0
408	**Yogurt**, fruit	150	580	630	60	330	5.8	Tr	4.0	330	Tr	Tr	Tr
409	*reconstituted*	43	170	180	17	94	1.7	0	1.1	94	0	0	0

[a] With high fortification, levels of Fe are approximately 16mg per 100g food

[b] With high fortification, levels of Fe are approximately 4mg per 100g food

No. 17-	Food	Retinol µg	Carotene µg	Vitamin[a] D µg	Vitamin[a] E mg	Thiamin[a] mg	Ribo-[a] flavin mg	Niacin[a] mg	Trypt 60 mg	Vitamin[a] B_6 mg	Vitamin B_{12} µg	Folate[a] µg	Panto- thenate mg	Biotin µg	Vitamin[a] C mg
	***Instant/granulated savouries** continued*														
400	**Vegetable based meal**	N[b]	N	7.4	3.70	0.90	0.50	4.8	2.5	0.60	1	74	2.60	7	44[c]
401	*reconstituted*	N[d]	N	1.9	0.93	0.23	0.13	1.2	0.6	0.15	Tr	19	0.65	2	11
	Instant/granulated sauces														
402	**Pour over sauce**, vegetable based	N[e]	N	5.9	4.50	1.40	0.28	2.3	N	0.17	1	56	2.10	7	36
403	*reconstituted*	N[f]	N	1.5	1.13	0.35	0.07	0.6	N	0.04	Tr	14	0.53	2	9
	Instant/granulated desserts														
404	**Cereal and milk based desserts**	N[b]	N	7.4	3.70	0.90	0.50	4.8	3.1	0.60	1	74	2.60	7	44[c]
405	*reconstituted*	N[d]	N	2.1	1.06	0.26	0.14	1.4	0.9	0.17	Tr	21	0.74	2	13
406	**Fruit desserts**	N[g]	N	7.2	3.50	0.90	0.40	4.7	2.4	0.60	1	72	2.40	6	43[c]
407	*reconstituted*	N[h]	N	2.1	1.00	0.26	0.11	1.3	0.7	0.17	Tr	21	0.69	2	12
408	**Yogurt**, fruit	N[i]	N	7.2	3.00	0.90	0.40	4.7	3.2	0.60	1	71	2.30	6	43[c]
409	*reconstituted*	N[j]	N	2.1	0.86	0.26	0.11	1.3	0.9	0.17	Tr	20	0.66	2	12

[a] Fortified values
[b] Vitamin A is 390µg per 100g food
[c] Levels ranged from 20mg to 60mg vitamin C per 100g
[d] Vitamin A is 98µg per 100g food
[e] Vitamin A is 338µg per 100g food
[f] Vitamin A is 85µg per 100g food
[g] Vitamin A is 380µg per 100g food
[h] Vitamin A is 109µg per 100g food
[i] Vitamin A is 370µg per 100g food
[j] Vitamin A is 106µg per 100g food

No. 17-	Food	Description and main data sources	Water g	Total Nitrogen g	Protein g	Fat g	Carbo-hydrate g	Energy value kcal	Energy value kJ
	Juices and juice drinks								
410	**Baby fruit juice**, fortified with vitamin C	10 samples, 2 brands; assorted flavours	90.1	0.02	0.1	Tr	8.0	30	130
411	**Baby fruit juice drink**, concentrated, fortified with vitamin C	10 samples of the same brand (Ribena)	33.5	0.04	0.3	Tr	57.7[ab]	218[b]	928[b]

[a] Includes 1.8g oligosaccharides
[b] Made up version contains 8.2g carbohydrate and provides abour 31 kcal and 131 kJ per 100g (1g product with 6g water)

Carbohydrate fractions and fatty acids, g per 100g food
Cholesterol, mg per 100g food

No. 17-	Food	Starch g	Total sugars g	Dietary fibre Southgate method g	Dietary fibre Englyst method g	Fatty acids Satd g	Fatty acids Mono-unsatd g	Fatty acids Poly-unsatd g	Cholest-erol mg
	Juices and juice drinks								
410	**Baby fruit juice**, fortified with vitamin C	0	8.0[a]	Tr	Tr	Tr	Tr	Tr	0
411	**Baby fruit juice drink**, concentrated, fortified with vitamin C	0	55.9[bc]	Tr	Tr	Tr	Tr	Tr	0

[a] Contains 2.2g glucose, 4.9g fructose and 0.9g sucrose per 100g food
[b] Contains 29.1g glucose, 18.8g fructose, 7.6g maltose and 0.4g sucrose per 100g food
[c] Made up version contains 8.0g total sugars per 100g

Inorganic constituents per 100g food

No. Food	mg										µg	
17-	Na	K	Ca	Mg	P	Fe	Cu	Zn	Cl	Mn	Se	I
Juices and juice drinks												
410 **Baby fruit juice**, fortified with vitamin C	4	170	11	5	6	0.2	Tr	Tr	3	0.04	Tr	Tr
411 **Baby fruit juice drink**, concentrated, fortified with vitamin C	2	250	7	4	5	Tr	Tr	Tr	Tr	Tr	Tr	Tr

No. 17-	Food	Retinol µg	Carotene µg	Vitamin D µg	Vitamin E mg	Thiamin mg	Riboflavin mg	Niacin mg	Trypt/60 mg	Vitamin B_6 mg	Vitamin B_{12} µg	Folate µg	Pantothenate mg	Biotin µg	Vitamin C mg
	Juices and juice drinks														
410	**Baby fruit juice**, fortified with vitamin C	0	N	0	N	Tr	Tr	0.1	0	0.01	0	0	0.05	0	54
411	**Baby fruit juice drink**, concentrated, fortified with vitamin C	0	N	0	N	0.01	0.01	0.1	0	0.02	0	4	0.06	Tr	440[a]

[a] Made up version contains 63mg vitamin C per 100g

DRIED INGREDIENTS

This section covers a number of dried vegetables and dried egg. These are mainly used as ingredients by food manufacturers. The values have been calculated from their raw counterparts. The varieties of vegetables presented here are assumed to be the same as those sold retail.

Composition of food per 100g

No. 17-	Food	Description and main data sources	Water g	Total Nitrogen g	Protein g	Fat g	Carbo-hydrate g	Energy value kcal	Energy value kJ
	Dried ingredients								
412	**Cabbage**, dried	Calculated from raw cabbage	10.5	2.46	15.4	3.6	38.0	237	1003
413	**Courgette**, dried	Calculated from raw courgette	10.5	4.09	25.6	5.7	25.5	249	1054
414	**Egg white**, dried	Calculated from raw egg white	4.0	11.80	73.8	Tr	Tr	295	1255
415	**Green beans**, dried	Calculated from raw green beans	10.5	2.93	18.3	4.8	30.8	232	982
416	**Mushroom**, dried	Calculated from raw mushroom	10.5	3.48[a]	21.8[b]	6.0	4.8	159	669
417	**Spinach**, dried	Calculated from raw spinach	10.5	3.89	24.3	6.9	13.8	211	889
418	**Sweetcorn**, dried	Calculated from raw sweetcorn	10.5	2.03	12.7	6.7	61.5[c]	342	1448

[a] Non protein nitrogen forms about 60% of total nitrogen

[b] (Total N - non protein nitrogen) x 6.25

[c] Includes oligosaccharides

Carbohydrate fractions and fatty acids, g per 100g food
Cholesterol, mg per 100g food

No. 17-	Food	Starch g	Total sugars g	Dietary fibre: Southgate method g	Dietary fibre: Englyst method g	Fatty acids: Satd g	Fatty acids: Mono-unsatd g	Fatty acids: Poly-unsatd g	Cholesterol mg
	Dried ingredients								
412	**Cabbage**, dried	0.9	37.1	(26.2)	(21.7)	(0.9)	Tr	(2.7)	0
413	**Courgette**, dried	1.4	24.1	N	(12.8)	(1.4)	Tr	(2.8)	0
414	**Egg white**, dried	0	Tr	0	0	Tr	Tr	Tr	0
415	**Green beans**, dried	8.7	22.1	(28.9)	(21.2)	(1.0)	Tr	(2.9)	0
416	**Mushroom**, dried	1.2	3.6	(27.8)	(13.3)	(1.2)	Tr	(3.6)	0
417	**Spinach**, dried	0.9	12.9	(33.9)	(18.2)	(0.9)	(0.9)	(4.3)	0
418	**Sweetcorn**, dried	54.4	7.1	(12.3)	(5.6)	(0.7)	(1.9)	(2.6)	0

Inorganic constituents per 100g food

No.	Food	mg										µg	
17-		Na	K	Ca	Mg	P	Fe	Cu	Zn	Cl	Mn	Se	I
	Dried ingredients												
412	**Cabbage**, dried	(45)	(2440)	(470)	(72)	(370)	(6.3)	(0.18)	(2.7)	(330)	(1.80)	(9)	(18)
413	**Courgette**, dried	(14)	(5110)	(360)	(310)	(640)	(11.4)	(0.28)	(4.3)	(640)	(1.40)	(14)	N
414	**Egg white**, dried	(1560)	(1230)	(41)	(90)	(270)	(0.8)	(0.16)	(0.8)	(1400)	Tr	(49)	(25)
415	**Green beans**, dried	Tr	(2210)	(350)	(160)	(370)	(11.5)	(0.10)	(1.9)	(87)	N	N	N
416	**Mushroom**, dried	(60)	(3870)	(73)	(110)	(970)	(7.3)	(8.71)	(4.8)	(840)	(1.20)	(110)	(36)
417	**Spinach**, dried	(1220)	(4350)	(1480)	(470)	(390)	(18.2)	(0.35)	(6.1)	(850)	(5.20)	(9)	(17)
418	**Sweetcorn**, dried	(4)	(970)	(11)	(140)	(340)	(2.6)	(0.15)	(1.5)	(41)	(0.70)	Tr	N

No. 17-	Food	Retinol µg	Carotene µg	Vitamin D µg	Vitamin E mg	Thiamin mg	Ribo-flavin mg	Niacin mg	Trypt/60 mg	Vitamin B_6 mg	Vitamin B_{12} µg	Folate µg	Panto-thenate mg	Biotin µg	Vitamin C mg
	Dried ingredients														
412	**Cabbage**, dried	0	N	0	(1.81)	Tr	Tr	(4.5)	N	Tr	0	Tr	Tr	Tr	Tr
413	**Courgette**, dried	0	N	0	N	Tr	Tr	(4.3)	N	Tr	0	Tr	Tr	Tr	Tr
414	**Egg white**, dried	0	0	0	0	0.08	3.50	0.7	21.7	0.16	1	105	2.50	57	0
415	**Green beans**, dried	0	N	0	(1.92)	Tr	Tr	(8.7)	N	Tr	0	Tr	Tr	Tr	Tr
416	**Mushroom**, dried	0	N	0	(14.86)	Tr	Tr	(38.7)	N	Tr	0	Tr	Tr	Tr	Tr
417	**Spinach**, dried	0	N	0	(14.86)	Tr	Tr	(10.4)	N	Tr	0	Tr	Tr	Tr	Tr
418	**Sweetcorn**, dried	0	N	0	(2.57)	Tr	Tr	(7.1)	N	Tr	0	Tr	Tr	Tr	Tr

Appendices

INDIVIDUAL FATTY ACIDS

The amounts of individual fatty acids are shown for selected fats and oils in grams per 100g food. The values for the unsaturated fatty acids include both the *cis* and *trans* fatty acids and the various positional isomers.

Names of fatty acids occurring in the tables

No of Carbon atoms and Double bonds	*Systematic name*	*Common name*
Saturated acids		
4:0	Butanoic acid	Butyric acid
6:0	Hexanoic acid	Caproic acid
8:0	Octanoic acid	Caprylic acid
10:0	Decanoic acid	Capric acid
11:0	Hendecanoic acid	
12:0	Dodecanoic acid	Lauric acid
14:0	Tetradecanoic acid	Myristic acid
15:0	Pentadecanoic acid	
16:0	Hexadecanoic acid	Palmitic acid
17:0	Heptadecanoic acid	Margaric acid
18:0	Octadecanoic acid	Stearic acid
19:0	Nonadecanoic acid	
20:0	Eicosanoic acid	Arachidic acid Arachic acid
22:0	Docosanoic acid	Behenic acid
24:0	Tetracosanoic acid	Lignoceric acid
Monounsaturated acids[a]		
10:1	Decenoic acid	
14:1	Tetradecenoic acid	Myristoleic acid
15:1	Pentadecenoic acid	
16:1	Hexadecenoic acid	Palmitoleic acid
17:1	Heptadecenoic acid	
18:1	Octadecenoic acid	Oleic acid
20:1	Eicosenoic acid	Eicosenic acid Gadoleic acid
22:1	Docosenoic acid	Erucic acid Cetoleic acid
24:1	Tetracosenoic acid	Nervonic acid Selacholeic acid

[a]Common names relate to the *cis* isomers

No of Carbon atoms and Double bonds	*Systematic name*	*Common name*
Polyunsaturated acids[a]		
16:2	Hexadecadienoic acid	
16:3	Hexadecatrienoic acid	
16:4	Hexadecatetraenoic acid	
18:2	Octadecadienoic acid	Linoleic acid
18:3	Octadecatrienoic acid	Linolenic acid[b]
18:4	Octadecatetraenoic acid	Stearidonic acid
20:2	Eicosadienoic acid	
20:3	Eicosatrienoic acid	Dihomolinolenic acid
20:4	Eicosatetraenoic acid	Arachidonic acid
• 20:5	Eicosapentaenoic acid (EPA)	
20:UNID	Unidentified C20 fatty acids	
21:5	Heneicosapentaenoic acid	
22:5	Docosapentaenoic acid	Clupanodonic acid
• 22:6	Docosahexaenoic acid (DHA)	Cervonic acid
22:UNID	Unidentified C22 fatty acids	

[a]Common names relate to all the *cis* isomers
[b] The main forms are alpha linolenic acid and gamma linolenic acid (GLA)

Fatty acids, g per 100g food

No. 17-	Food	Saturated 4:0	6:0	8:0	10:0	12:0	14:0	15:0	16:0	17:0	18:0	20:0	22:0	24:0
	Cooking fats													
2	**Cocoa butter**	0	0	0	0	0	0	0	24.7	0	33.3	1.0	0	0
3	**Cocoa butter alternative**[a]	0	0	1.0	1.0	17.2	6.4	0	16.2	0	17.2	0.3	0	0
4	**Compound cooking fat**	0	0	0	0.1	0.2	9.0	1.1	24.8	1.3	9.2	2.2	1.6	0
5	polyunsaturated	0	0	0	0	Tr	0.3	Tr	10.6	0.1	8.6	0.4	0.6	0
6	**Dripping**, beef	0	0	0	0	0.1	3.2	0.6	25.0	1.7	21.4	0.5	0	0
9	**Ghee**, vegetable	0	0	0.2	0	0.2	1.0	0	41.7	0.1	4.8	0.4	0.1	0
10	**Lard**[b]	0	0	0	Tr	Tr	1.4	0	24.4	0.3	14.1	0	0	0
11	**Suet**, shredded	0	0	0	0	0.1	2.7	0.4	23.6	1.3	21.5	0.2	0	0
12	vegetable	0	0	0	0	0.2	0.7	0	27.2	0.1	16.0	0.4	0.2	0.1
	Spreading fats													
13	**Butter**	3.0	2.0	1.2	2.6	3.1	8.7	1.4	21.9	1.3	9.2	0.3	0	0
14	spreadable[c]	(3.0)	1.5	1.1	2.7	3.5	8.6	2.4	19.6	1.3	7.3	0.1	0	0
15	**Blended spread** (70-80% fat)	0.9	0.5	0.2	0.5	0.8	2.6	0.5	13.2	0.4	5.5	0.3	0.1	0
16	(40% fat)	0.8	0.5	0.2	0.5	0.5	1.8	0.3	8.5	0.2	4.7	0.1	0	0
18	**Margarine**, hard, animal and vegetable fats	0	0	0	0.1	0.2	6.2	0.8	17.9	0.9	6.3	1.3	1.0	0
19	-, vegetable fats	0	0	0.6	0.6	8.8	3.4	15.3	0	0.1	6.3	0.4	0.4	0
20	soft, not polyunsaturated	0	0	0	Tr	0.2	5.0	0.5	14.7	0.7	4.4	1.0	0.7	0
21	-, polyunsaturated	0	0	0	0	0.1	0.2	Tr	10.1	0.1	5.6	0.5	0.4	0.1

[a] Contains 0.1g other unidentified saturated fatty acids per 100g food
[b] Contains 0.1g 19:0 per 100g food
[c] Contains 0.2g 13:0 per 100g food

No.	Food	Monounsaturated							Polyunsaturated			
		14:1	15:1	16:1	17:1	18:1	20:1	22:1	18:2	18:3	20:2	22:6
17-												
	Cooking fats											
2	**Cocoa butter**	0	0	0.5	0	32.3	0	0	2.9	0.5	0	0
3	**Cocoa butter alternative**	0	0	0	0	32.4	0	0	2.9	0	0	0
4	**Compound cooking fat**	Tr	0.1	9.3	0.2	23.7	4.6	3.3	4.6	0.7	0	0
5	polyunsaturated	0	0	0.2	Tr	30.8	0.2	0	43.4	0.5	0	0
6	**Dripping**, beef	1.0	0	2.3	0.8	34.8	0	0	1.8	0.7	0	0
9	**Ghee**, vegetable	0	0	0.2	0	36.7	0.1	0	9.4	0.3	0	0
10	**Lard**[a]	0	0	2.4	0.3	39.8	0.9	0	8.7	0.6	0	0
11	**Suet**, shredded[b]	0.7	0	1.3	0.5	27.9	0	0	1.7	0.4	0	0
12	vegetable	0	0	0.1	0	26.2	0.1	0	12.5	0.2	0	0
	Spreading fats											
13	**Butter**[c]	0.7	0	1.2	0.3	17.7	0	0.1	1.8	0.9	0	0
14	spreadable[d]	0	0	1.7	0.7	22.5	1.0	0.2	2.8	0.9	0	0
15	**Blended spread** (70-80% fat)[e]	0.2	Tr	0.4	0.1	36.0	0.6	0.1	7.3	1.2	0	0
16	(40% fat)	0.2	0	0.2	0	12.6	0.1	0.3	6.5	0.8	0	0
18	**Margarine**, hard, animal and vegetable fats	0.1	0	6.1	0.2	19.6	6.8	3.4	4.7	0.6	0.1	0
19	-, vegetable fats	0.2	0	0.4	0	31.8	0.4	0.2	8.0	1.2	0	0
20	soft, not polyunsaturated	Tr	Tr	5.1	0.2	29.1	3.2	1.2	9.7	2.3	0.5	0
21	-, polyunsaturated	0	0	0.2	Tr	26.3	0.2	0	33.8	2.1	0	0

[a] Contains 0.4g 20:4 and 0.3g 22:2 per 100g food
[b] Contains 0.1g 20:UNID per 100g food
[c] Contains 0.1g 16:2 per 100g food
[d] Contains 0.3g 20:UNID, 0.4g 22:UNID per 100g food
[e] Contains 0.1g 10:1 per 100g food

Fatty acids, g per 100g food

No.	Food	Saturated 4:0	6:0	8:0	10:0	12:0	14:0	15:0	16:0	17:0	18:0	20:0	22:0	24:0
17-														
22	**Fat spread** (70-80% fat), not polyunsaturated	0	0	0	0.1	0.3	2.5	0.5	15.8	1.2	9.9	0.1	0	0
23	(70% fat), polyunsaturated	0	0	0	0	0	0.1	0	5.9	0.1	5.6	0.3	0.5	0.1
24	(60% fat), polyunsaturated	0	0	0	0	0.1	0.2	Tr	7.2	0.1	3.4	0.3	0	0
25	-, with olive oil	0	0	0	0	0.1	0.2	0	7.1	0.1	3.3	0.3	0.2	0
26	(40% fat), not polyunsaturated	0	0	0	0.1	1.2	0.6	Tr	6.7	Tr	2.8	0.2	0.1	0
28	(20-25% fat), not polyunsaturated	0	0	0	Tr	0.1	0.2	Tr	4.2	Tr	1.5	0.1	0.1	0
Oils														
31	**Coconut oil**	0	0.4	6.9	6.2	45.0	17.0	0	8.4	0	2.5	0.1	0	0
32	**Cod liver oil**	0	0	0	0	0	5.6	Tr	12.5	0	2.0	Tr	0	0
33	**Corn oil**	0	0	0	0	0.1	0.1	0	11.3	0	2.1	0.5	0.2	0.2
34	**Cottonseed oil**	0	0	0	0	Tr	0.8	0	22.4	0	2.5	0.3	0.1	0
35	**Evening primrose oil**	0	0	0	0	0	Tr	0	5.6	0.1	1.7	0.3	0.1	0
36	**Grapeseed oil**	0	0	0	0	0	Tr	0	6.8	0.1	3.9	0.2	0.1	Tr
37	**Hazelnut oil**	0	0	0	0	0	Tr	0	5.3	0.1	2.3	0.1	0	0
38	**Olive oil**	0	0	0	0	0.1	0.1	0	10.1	0.1	3.0	0.4	0.1	0.4
39	**Palm oil**	0	0	0	0	0.1	1.0	0	41.8	0	4.6	0.3	0	0
40	**Peanut oil**	0	0	0	0	Tr	Tr	0	10.9	0	3.2	1.3	3.2	1.4
41	**Rapeseed oil**	0	0	0	0	0	Tr	0	4.2	Tr	1.5	0.6	0.3	Tr
42	**Safflower oil**	0	0	0	0	0	0.1	0	6.6	0	2.3	0.3	0.3	0.1

No.	Food	Monounsaturated 14:1	15:1	16:1	17:1	18:1	20:1	22.1	Polyunsaturated 18:2	18:3	20:2	22:6
17-												
22	**Fat spread** (70-80% fat), not polyunsaturated	0.3	Tr	2.2	0.1	28.2	0.4	0	5.5	1.0	0.1	0
23	(70% fat), polyunsaturated	0	0	0.1	0	20.9	0.1	0	32.5	0.5	0	0
24	(60% fat), polyunsaturated	0	0	0.1	Tr	17.9	0.1	0	28.5	0.1	0	0
25	-, with olive oil[a]	0	0	0.2	0.1	35.6	0.3	0.1	10.1	2.3	0	0
26	(40% fat), not polyunsaturated	Tr	0	0.1	0	20.0	0.3	0.2	5.1	1.3	0	0
28	(20-25% fat), not polyunsaturated	0	0	0.1	Tr	11.6	0.1	0.1	4.2	0.7	0	0
Oils												
31	**Coconut oil**	0	0	0	0	6.0	0	0	1.5	0	0	0
32	**Cod liver oil**[b]	0	0	7.1	18.3	0	11.1	8.1	2.6	1.1	0	8.3
33	**Corn oil**	0	0	0.2	0	29.4	0.3	Tr	50.4	0.9	0	0
34	**Cottonseed oil**	0	0	0.8	0	17.4	0.1	Tr	50.1	0.1	Tr	0
35	**Evening primrose oil**	0	0	Tr	Tr	10.4	0.2	Tr	68.4	8.2	0	0
36	**Grapeseed oil**	0	0	0.1	Tr	15.4	0.2	0.1	67.8	0.4	0	0
37	**Hazelnut oil**	0	0	0.2	0.1	76.1	0.1	0	11.1	0.1	0	0
38	**Olive oil**	0	0	0.7	0.1	71.9	0.3	0	7.5	0.7	0	0
39	**Palm oil**	0	0	Tr	0	37.1	0	0	10.1	0.3	0	0
40	**Peanut oil**	0	0	Tr	0	43.3	1.0	0.1	31.0	0	0	0
41	**Rapeseed oil**	0	0	0.2	Tr	57.6	1.2	0.2	19.7	9.6	Tr	0
42	**Safflower oil**[a]	0	0	0.1	0	11.4	0.2	0.2	73.9	0.1	0	0

[a] Contains 0.1g 24:1 per 100g food
[b] Contains 0.6g 16:2, 0.4g 16:3, 1.0g 16:4, 2.1g 18:4, 0.9g 20:4, 10.8g 20:5, 0.7g 21:5 and 1.4g 22:5, per 100g food

Fatty acids, g per 100g food

No.	Food	Saturated 4:0	6:0	8:0	10:0	12:0	14:0	15:0	16:0	17:0	18:0	20:0	22:0	24:0
17-														
43	**Sesame oil**	0	0	0	0	0	Tr	0	8.6	0.1	5.1	0.6	0.1	0.1
44	**Soya oil**	0	0	0	0	0	0.1	0	10.7	0	3.8	0.4	0.5	0.1
45	**Sunflower oil**	0	0	0	0	0	0.1	0	6.2	0	4.3	0.3	0.8	0.3
47	**Walnut oil**	0	0	0	0	0	Tr	0	6.5	0.1	2.4	0.1	Tr	0
48	**Wheatgerm oil**	0	0	0	0	Tr	Tr	0	17.2	0.1	0.9	0.1	0.1	0.1

No.	Food	Monounsaturated 14:1	15:1	16:1	17:1	18:1	20:1	22.1	Polyunsaturated 18:2	18:3	20:2	22:6
17-												
43	**Sesame oil**	0	0	0.1	Tr	37.2	0.2	0	43.1	0.3	0	0
44	**Soya oil**	0	0	0.1	0	20.8	0.2	0.1	51.5	7.3	0	0
45	**Sunflower oil**	0	0	0.1	0	20.2	0.1	0.1	63.2	0.1	0	0
47	**Walnut oil**	0.1	0	0	0.1	16.2	0	0	58.4	11.5	0	0
48	**Wheatgerm oil**	0	0	0.2	Tr	15.4	1.0	0.1	55.1	5.3	0	0

VITAMIN E FRACTIONS

The vitamin E activity of foods can be derived from a number of different tocopherols and tocotrienols. Where vitamin E is present, and the amount of each tocopherol was known, the values are shown below. The total vitamin E activity is also shown as α-tocopherol equivalents, which has been taken as the sum of the α-tocopherol, 40% of the β-tocopherol, 10% of the γ-tocopherol, 1% of δ-tocopherol, 30% of α-tocotrienol, 5% of the β- tocotrienol, 1% of the γ-tocotrienol and 1% of δ-tocotrienol (McClaughlin and Weihrauch 1979)[a]

Vitamin E fractions, mg per 100g food

No. *17-*	Food	α- tocopherol	β– tocopherol	γ– tocopherol	δ– tocopherol	Vitamin E equiv
	Cooking fats					
2	**Cocoa butter**	0.50	0.16	4.90	0	1.1
12	**Suet**, vegetable	17.97	0	0	0	18.0
	Spreading fats					
14	**Butter**, spreadable	2.92	Tr	Tr	Tr	2.9
15	**Blended spread** (70-80% fat)	10.08	0.07	11.64	1.22	11.3
16	(40% fat)	2.24	0.26	14.86	4.76	3.9
18	**Margarine**, hard, animal and vegetable fats	4.28	0	1.56	0.27	4.4
20	soft, not polyunsaturated	11.59	Tr	7.40	0.54	12.3
21	-, polyunsaturated	31.15	1.29	8.95	3.46	32.6
22	**Fat spread** (70-80% fat), not polyunsaturated	2.22	0.07	2.17	6.04	2.5
24	(60% fat), polyunsaturated	29.81	1.20	4.43	1.27	30.8
26	(40% fat), not polyunsaturated	7.27	0	7.29	0.42	8.0
28	(20-25% fat), not polyunsaturated	4.71	0	3.98	0.45	5.1

[a] McClaughlin, T.J. and Weihrauch, J.L. (1979). Vitamin E content of foods. *J Am. Diet. Assoc.* **76**, 647-665

No. 17-	Food	α-tocopherol	β–tocopherol	γ–tocopherol	δ–tocopherol	Vitamin E equiv
	Oils					
31	**Coconut oil**	0.50	0	0	0.60	0.7[a]
33	**Corn oil**	11.20	0	60.20	1.80	17.2
34	**Cottonseed oil**	38.90	0	38.70	0	42.8
38	**Olive oil**	5.10	0	Tr	0	5.1
39	**Palm oil**	25.60	0	31.60	7.00	33.1[b]
40	**Peanut oil**	13.0	0	21.4	2.1	15.2
41	**Rapeseed oil**	18.40	0	38.0	1.20	22.2
42	**Safflower oil**	38.70	0	17.40	24.00	40.7
44	**Soya oil**	10.10	0	59.30	26.40	16.1
45	**Sunflower oil**	48.70	0	5.10	0.80	49.2
48	**Wheatgerm oil**	133.0	0	26.0	27.10	136.7[c]
	Chocolate confectionery					
83	**Chocolate covered caramels**	2.03	0.08	3.14	0.19	2.37
84	**Chocolate covered bar with fruit/nuts and wafer/biscuit**	2.51	0.13	3.46	0.37	2.91
88	**Chocolate**, fancy and filled	1.35	0.05	2.73	0.31	1.65
89	milk	0.38	0.02	0.66	0.06	0.45
90	plain	0.86	0.18	5.11	0.28	1.44
91	white	0.61	Tr	5.26	0.21	1.14
93	**Kit Kat**	0.58	0.04	4.06	0.22	1.03[d]
95	**Milky Way**	1.36	0.03	2.10	0.13	1.91[e]
96	**Smartie-type sweets**	0.46	Tr	3.20	0.18	0.80[f]
100	**Twix**	2.74	0.09	2.74	0.18	3.72[g]

[a] Includes contribution from 0.50mg α-tocotrienol
[b] Includes contribution from 14.3mg α-tocotrienol
[c] Includes contribution from 2.60mg α-tocotrienol
[d] Includes contribution from 0.08mg α-tocotrienol, 0.13mg γ-tocotrienol
[e] Includes contribution from 1.04mg α-tocotrienol, 1.2mg γ-tocotrienol, 0.23mg δ-tocotrienol
[f] Includes 0.07mg α-tocotrienol, 0.06mg γ-tocotrienol
[g] Includes contribution from 2.13mg α-tocotrienol, 2.35mg γ-tocotrienol, 0.52mg δ-tocotrienol

No. 17-	Food	α–tocopherol	β–tocopherol	γ–tocopherol	δ–tocopherol	Vitamin E equiv
	Non chocolate confectionery					
103	**Cereal crunchy bar**	3.37	0.16	4.07	0.35	3.84
104	**Chew sweets**	0.86	0.04	0.38	0.11	0.91
106	**Foam sweets**	0.02	Tr	0.02	Tr	0.02
	Savoury snacks					
123	**Breadsticks**	0.32	0.14	0.43	0.07	0.44
125	**Corn snacks**	5.38	0.11	4.09	0.54	5.80
133	**Potato crisps**	5.42	0.22	2.88	0.73	5.83
136	low fat	3.36	0.09	0.62	0.06	3.47
146	**Punjabi puri**	4.54	0.24	8.72	0.71	5.52
149	**Tortilla chips**	1.72	0.06	1.97	0.15	1.94
150	**Twiglets**	2.02	0.48	2.56	0.19	2.47
151	**Wheat crunchies**	2.50	0.21	0.98	Tr	2.68
	Liqueurs					
242	**Cream liqueurs**	0.56	0	0.05	0	0.57
	Sauces					
316	**Mayonnaise**	14.80	Tr	40.80	11.30	18.99
318	reduced calorie	7.74	0.06	5.66	0.44	8.33
	Miscellaneous foods					
375	**Tomatoes**, sun dried	23.19	1.19	3.17	0.23	23.98

PHYTOSTEROLS

Plants contain a number of phytosterols (plant sterols) which are distinct from cholesterol. In plant oils, the three most common sterols are β-sitosterol, campesterol and stigmasterol. There may also be measurable amounts of at least nine other phytosterols.

The amounts of the six main phytosterols are shown below for selected fats and oils.

Phytosterols, mg per 100g food

No. 17-	Food	β– sito-sterol	Campe-sterol	Stigma-sterol	Δ^{7}-Stigma-stenol	Brassica-sterol	Δ^{5}-Avena-sterol
	Cooking fats						
2	**Cocoa butter**	138	22	61	0	Tr	0
9	**Ghee**, vegetable	33	11	7	0	0	0
	Spreading fats						
13	**Butter**	4	0	0	0	0	0
14	spreadable	Tr	Tr	Tr	Tr	Tr	Tr
15	**Blended spread** (70-80% fat)	136	105	3	1	18	5
16	(40% fat)	38	14	10	Tr	Tr	Tr
18	**Margarine**, hard, animal and vegetable fats	35	27	1	2	3	1
19	-, vegetable fats[a]	130	52	13	0	9	0
20	soft, not polyunsaturated	95	68	5	2	15	3
21	-, polyunsaturated	139	43	33	15	1	10[b]
22	**Fat spread** (70-80% fat), not polyunsaturated	23	15	1	Tr	3	1
23	(70% fat), polyunsaturated	184	68	10	0	Tr	Tr

[a] Contains 16mg other phytosterols per 100g food
[b] Includes 6mg Δ^{7}-avenasterol

No. 17-	Food	β– sitosterol	Campesterol	Stigmasterol	Δ^{7}-Stigmastenol	Brassicasterol	Δ^{5}-Avenasterol
	***Spreading fats** continued*						
24	**Fat spread** (60% fat), polyunsaturated	122	20	16	29	1	10[a]
25	(60% fat), with olive oil	147	64	12	0	Tr	Tr
26	(40% fat), not polyunsaturated	94	68	2	0	14	3
28	(20-25% fat), not polyunsaturated	60	41	2	Tr	8	2
	Oils						
31	**Coconut oil**	66	7	15	0	Tr	0
33	**Corn oil**	595	179	51	0	0	20
34	**Cottonseed oil**	303	20	0	0	Tr	5
36	**Grapeseed oil**	98	13	15	0	0	0
38	**Olive oil**	102	2	1	0	0	9
39	**Palm oil**	28	7	4	0	0	0
40	**Peanut oil**	179	29	26	0	0	0
41	**Rapeseed oil**	129	95	4	0	23	0
42	**Safflower oil**	231	49	40	0	0	5
43	**Sesame oil**	430	164	60	0	0	0
44	**Soya oil**	194	65	71	0	0	0
45	**Sunflower oil**	210	32	35	0	Tr	14
47	**Walnut oil**	155	9	0	0	0	0
48	**Wheatgerm oil**	370	122	Tr	0	0	33

[a] Includes 6mg Δ^{7}-avenasterol

PERCENTAGE OF ALCOHOL BY VOLUME IN SELECTED BEERS

In the main tables, the amount of alcohol in each alcoholic beverage is given as grams per 100ml. However, most alcoholic beverages as bought now show their strength in terms of alcohol *by volume* (ABV) since it is a legal requirement for this information to be given if the product contains 1.2 per cent alcohol or more. Drinks with lower strength must also show their ABV if a claim is made about their alcoholic content. The relationship between the two presentations can be derived from the specific gravity of alcohol, which is 0.79; thus a drink containing 10 per cent alcohol by volume contains 7.9 grams of alcohol per 100ml.

The amount of alcohol in most types of beer can vary substantially according to brand and exact description. The values in the main tables are averages, but typical ranges (as percent ABV, not g/100ml) for a number of beers as given below.

Percent alcohol by volume in selected beers

	Mean	Typical range
Ales		
Brown ale, bottled	3.2	2.6 - 4.7
cask conditioned	4.9	2.6 - 9.2
export	4.4	3.9 - 6.0
keg	3.4	2.7 - 4.0
Light ale	3.3	2.9 - 3.7
Mild ale	3.3	3.0 - 3.5
Old ale	6.4	2.6 - 9.4
Pale ale	3.6	2.7 - 4.7
keg	3.3	3.0 - 4.2
strong	6.7	4.4 - 7.2
Strong ale	5.4	4.2 - 12.5
Bitter		
Bitter	3.7	3.0 - 6.0
best	3.9	3.3 - 5.0
cask conditioned	4.1	3.2 - 5.7
keg	3.9	3.2 - 5.0
Mild	3.2	2.6 - 3.9
dark	3.2	2.9 - 3.5
keg	3.2	3.0 - 3.6
premium	4.3	4.1 - 4.5
Premium	4.4	4.0 - 5.0

	Mean	Typical range
Lager		
Lager	3.8	3.0 - 5.0
extra	4.6	3.0 - 6.0
Pilsner	5.2	3.3 - 8.0
Premium	4.9	4.0 - 6.3
strong	6.0	5.0 - 8.4
very strong	9.1	8.9 - 9.5
Stout		
Stout	4.0	2.8 - 5.0
sweet	3.1	2.0 - 3.8
extra strong	8.9	7.7 - 10.0

Alcohol-free beer contains less than 0.05% ABV

Low alcohol beer contains less than 1.2% ABV

Source: Adapted from The Brewery Digest Manual and Who's Who in British Brewing and Scottish Whisky Distilling (1993) Hampton Publishing Limited, England

MISCELLANEOUS INGREDIENTS AND ADDITIVES

A number of food ingredients and additives consist of or contain nitrogen which is not necessarily protein, lipid matter which is not necessarily fat, or carbohydrates which may or may not be absorbed. They may also provide energy because of their alcohol, organic acids, polyols or other energy-yielding constituents. Most are used in such small amounts in foods that these and their energy value can be ignored for nutritional purposes, but some are used in larger amounts. The actual amount of energy that they yield will vary, but for nutrition labelling purposes the Food Labelling (Amendment) Regulations 1994 state that all organic acids provide 3 kcal (13 kJ) per gram; all polyols provide 2.4 kcal (10 kJ) per gram; that protein, defined as total nitrogen x 6.25, and all metabolisable carbohydrates provide 4 kcal (17 kJ) per gram; and that all lipids provide 9 kcal (37 kJ) per gram. The following table gives nutrient values for the pure substances that reflect these rules. In addition, polydextrose is provisionally suggested as providing 1 kcal (5 kJ) per gram.

Composition per 100g

	Protein	Fat	Carbo-hydrate[a]	Energy value	
	g	g	g	kcal	kJ
Proteins and amino acids					
Amino acids and their salts	100	0	0	400	1700
Collagen	100	0	0	400	1700
Hydrolysed protein	100	0	0	400	1700
Lipids					
Fatty acid salts	0	100	0	900	3700
Lecithin	0	100	0	900	3700
Mono- and di-glycerides	0	100	0	900	3700
Carbohydrates					
Cellulose and modified celluloses	0	0	0	0	0
Dextrins	0	0	100	400	1700
Gums (e.g. agar, alginates, carrageenan, gellan gum, guar gum, locust bean (carob) gum, xanthan gum)[b]	0	0	0	0	0
Pectin[b]	0	0	0	0	0
Polydextrose	0	0	100	100	500
Polyols (e.g. glycerol, isomalt, lactitol, maltitol, mannitol, sorbitol, xylitol)	0	0	100	240	1000
Starch and modified starches	0	0	100	400	1700

[a] Expressed as total carbohydrate and not as monosaccharide equivalent

[b] These polysaccharides can be partly fermented in the large intestine to short chain fatty acids which provide some energy, but this energy is at present discounted

	Protein	Fat	Carbo-hydrate[a]	Energy value	
	g	g	g	kcal	kJ
Miscellaneous					
Ethanol[b]	0	0	0	700	2900
Organic acids (e.g. acetic, citric, fumaric, gluconic, malic, succinic and tartaric) and their salts	0	0	0	300	1300

[a] Expressed as total carbohydrate and not as monosaccharide equivalent

[b] The energy value of 100ml ethanol is 553 kcal or 2290 kJ

Unless specified, the recipes use whole pasteurised milk, fresh cream, Cheddar cheese and plain white flour.

An average egg has been assumed to weigh 50g, an average egg yolk 18g and an average egg white 32g. A level teaspoon refers to a standard 5 ml spoon and has been taken to hold 5g salt and 3g spices.

The type of fat used in recipes has been specified. The vegetable oil was a retail blended vegetable oil. Margarine was an average of hard, soft and polyunsaturated types. The butter was salted.

For fried dishes, the fat absorbed during frying has been included at the end of the ingredients list with the quantity absorbed shown in brackets.

54 Icing, butter

125g butter
210g icing sugar
15ml sherry

Beat butter until light and creamy. Gradually beat in half the icing sugar. Beat in sherry gradually with remaining icing sugar.

55 Icing, fondant

10ml lemon juice
2 egg whites
2 tsp liquid glucose
450g sugar
colouring
flavouring

Put ingredients in a bowl and beat to make a pliable smooth mixture. Add colouring and flavouring and knead.

56 Icing, glacé

250g icing sugar
3 tbsp water

Place the sugar into a bowl, stir in the water until the icing is the same consistency as thick cream.

57 Icing, Royal

2 egg whites
560g icing sugar
5ml lemon juice

Beat egg whites until light and fluffy. Gradually beat in icing sugar until mixture is stiff and stands in peaks. Blend in lemon juice.

77 **Lemon curd, homemade**

100g butter
150ml lemon juice
300g sugar
4 eggs

Melt butter, lemon juice and all the sugar in a pan. Add the eggs one by one and cook slowly, stirring all the time, until the mixture coats the back of the spoon. Pour into jars and cover.

Weight loss: 17%

81 **Mincemeat, vegetarian**

100g apple, peeled, cored
100g currants
100g sultanas
100g Demerara sugar
100g vegetable suet
25g mixed spice
25g rind and juice of 1 lemon
25g rind and juice of 1 orange
100g mixed peel
70ml brandy
70ml sherry

Put the apples, currants and sultanas through a food processor or chop. Turn into a mixing bowl and add sugar, vegetable suet, spices, the rind and juice of the lemon and orange, mixed peel and the spirits. Stir well. Cover and leave for 2 days. Pot, cover and store as for jam.

Weight loss: 8%

98 **Truffles, mocha**

225g plain chocolate
5g coffee, instant
10ml water
60ml condensed milk
100g grated chocolate

Break 125g of the chocolate into pieces and melt in a bowl over hot water, making sure the bottom of the bowl does not touch the water. Dissolve coffee in warm water and stir into the melted chocolate along with the condensed milk. Allow the mixture to cool slightly. Form into small balls and roll in grated chocolate.

Weight loss: 8%

99 **Truffles, rum**

225g plain chocolate
2 egg yolks
25g butter
10ml rum
15ml single cream
100g grated plain chocolate

Break 125g of the chocolate into pieces and melt the chocolate in a bowl over hot water, making sure the bottom of the bowl does not touch the water. Add the egg yolks, butter, rum and cream and stir. Allow the mixture to cool slightly. Form into balls and roll in grated chocolate.

Weight loss: 7%

105 **Coconut ice**

454g granulated sugar
142ml millk
142g desiccated coconut
few drops of vanilla essence
few drops of pink food colouring

Dissolve sugar in milk, boil gently for 10 minutes. Remove from heat, stir in coconut and vanilla essence. Pour half into tin, colour other half and pour over white mixture. Mark into bars when almost set.

Weight loss: 1%

109 **Fudge**

450g granulated sugar
175ml evaporated milk
150ml milk
75g butter
few drops of vanilla essence

Dissolve sugar in milks and add butter. Bring to the boil and boil gently to 125°C. Remove from heat, add vanilla essence. Beat mixture until thick and grainy. Pour into tin and cut into squares when almost set.

Weight loss: 28%

110 **Halva, carrot**

800ml milk
2g cardamom, crushed
900g carrots, grated
100g vegetable ghee
225g granulated sugar
25g pistachio nuts, chopped

Bring the milk to the boil in a heavy based pan. Add the cardamoms and carrots and simmer for 1-1½ hours until thick. Add ghee and fry for 2 minutes. Add sugar and cook gently for 5-10 minutes. Add pistachio nuts and stir.

Weight loss: 62%

111 **Halva, semolina**

550g granulated sugar
900ml water
½ tsp saffron
225g vegetable ghee
225g semolina
2g cardamoms, crushed
25g pistachio nuts, chopped
25g almonds

In a pan bring sugar, water and saffron to the boil then allow to cool. Heat ghee in a heavy based saucepan and stir in semolina over a gentle heat. Cook for 10 minutes. Add sugar water a little at a time, stirring continuously. Add cardamoms, pistachio nuts and almonds.

Weight loss: 21%

115 **Nougat**

450g caster sugar
150ml water
½ tsp cream of tartar
60g honey
4 egg whites
100g pistachio nuts, blanched and chopped
few drops of vanilla essence
rice paper for lining tin

Put sugar and water in a saucepan and heat until sugar has dissolved and mixture reaches 132°C. Stir in cream of tartar. Put honey into another saucepan and heat until very hot. Whisk egg whites until stiff. Put the egg whites over a saucepan of boiling water; gradually pour over sugar mixture, stirring well. When thoroughly blended, add hot honey, chopped nuts and vanilla essence. Stir the mixture for a few minutes, the temperature should be around 121°C. Pour mixture into a 9-10inch square tin lined with rice paper. Allow to set. Cut into neat pieces and wrap in waxed paper.

Weight loss: 26%

116 **Peanut brittle**

400g granulated sugar
175g brown sugar
175g golden syrup
150ml water
50g butter
1g bicarbonate of soda
350g unsalted peanuts, chopped

Put sugars, syrup and water in a pan and heat gently until sugar has dissolved. Add butter, bring to the boil very gently to 149°C. Stir in bicarbonate of soda and nuts. Pour into tin and mark into bars when almost set.

Weight loss: 18%

118 **Peppermint creams**

450g icing sugar
5ml lemon juice
1 egg white
1 tsp peppermint essence

Mix the sugar, lemon juice, egg white and peppermint essence to make a pliable mixture. Knead and roll out to 0.5cm (¼ inch) thick. Cut into rounds or form into balls.

121 **Turkish delight, with nuts**

25g gelatine
300ml water
450g granulated sugar
¼ tsp lemon juice
30g pistachio nuts
½ tsp rose water
few drops of pink food colouring
50g icing sugar
1g bicarbonate of soda

Dissolve gelatine in water over a gentle heat. Add sugar and lemon juice. Stir gently until the sugar dissolves. Remove from heat and allow to cool for 10 minutes. Stir in pistachio nuts, rose water and food colouring. Pour into tin. Leave to set for 24 hours. Cut and roll in icing sugar and bicarbonate of soda.

Weight loss: 25%

124 **Chevda/chevra/chewra**

17g lentils
16ml water
35g peanuts, plain
18g rice flakes
8g vegetable oil
2g mixed spice
2g chilli powder
2g salt

Soak lentils in water overnight, drain and dry. Fry lentils, peanuts and rice flakes separately in vegetable oil. Mix the cooked ingredients with the mixed spices, chilli powder and salt.

Weight loss: 20%

130 **Popcorn, candied**

45ml vegetable oil
75g popping corn

glaze
45ml water
200g caster sugar
25g butter

Prepare corn as for plain popcorn (No. 131). Heat glaze ingredients until sugar has dissolved, boil to soft ball stage (125°C). Add the popped corn and stir until coated.

Weight loss: 7% (popcorn), 15.2% (glaze)

131 **Popcorn, plain**

45ml vegetable oil
75g popping corn

Heat oil gently in saucepan until test corn pops. Remove from heat, add corn, cover and return to heat until all corn has popped.

Weight loss: 7%

148 **Sev/ganthia**

220g besan flour
4g salt
105ml water
(55g vegetable oil)

Mix flour, salt and water to make a thick batter. Force through a sev/siawa machine into hot oil and deep fry until golden brown.

Weight loss: 30%

164 **Coffee, Irish**

30ml whisky
1 tsp brown sugar
120ml black coffee
30ml double cream

Add whisky and sugar to a warmed glass. Pour in black coffee and stir to dissolve sugar. Add cream poured over the back of a spoon.

199 **Lemonade, homemade**

15g lemon rind
175g caster sugar
900ml boiling water
60ml lemon juice

Put lemon rind and sugar into a bowl and pour on boiling water. Cover and leave to cool, stirring occasionally. Add lemon juice and strain. Serve chilled.

Weight loss: 4%

227 **Mulled wine, homemade**

400g orange, sliced
5g cloves
1500ml port
225g caster sugar
5g cinnamon
2g nutmeg
2g mace
160g lemon, sliced

Add orange, cloves, port, sugar and remaining spices to a saucepan. Slowly heat until a white foam appears. Do not boil. Stir well to dissolve sugar. Remove from heat. Slice lemons thinly and add to the wine. Strain before serving.

Weight loss: 5%

243 **Egg nog**

1 egg
25g caster sugar
50ml rum
300ml milk

Whisk egg and sugar together, stir in alcohol. Heat the milk without boiling and pour over the egg mixture, stirring well.

Weight loss: 2%

248 Bouillabaisse

½g saffron
140ml boiling water
140g olive oil
300g onion, sliced
30g celery, chopped
225g tomatoes, canned
6g garlic clove, crushed
2 bay leaves
½ tsp thyme
½ tsp salt
½ tsp pepper
180g John Dory, cubed
180g monkfish, cubed
180g red mullet, cubed
180g bass, cubed
180g prawns

Put saffron in a bowl with boiling water and leave for 30 minutes. Heat oil in a saucepan, add onions and celery and cook for 5 minutes. Stir in tomatoes, garlic, herbs and seasoning. Add fish and pour over saffron liquid and enough water to cover the fish. Bring to the boil and simmer for 8 minutes. Add prawns and cook for 5 minutes.

Weight loss: 7%

249 Carrot and orange soup

500g carrot, chopped
1000ml water
100g onion, chopped
60ml orange juice
½ tsp salt
½ tsp pepper
5g coriander, chopped
20ml single cream

Put ingredients into a saucepan and bring to the boil. Simmer for 30 minutes. Sieve or liquidise and return to the pan to reheat. Add seasoning and garnish with coriander and cream.

Weight loss: 19%

256 French onion soup

900g onion, sliced
5g sugar
30ml vegetable oil
750ml beef stock
3g salt
1g pepper

Fry onions and the sugar in the oil until soft. Add stock and bring to the boil. Simmer for 15 minutes and adjust seasoning.

Weight loss: 11%

257 Gazpacho

425g tomatoes, canned
250ml vegetable stock
75g olive oil
60ml vinegar
140g tomatoes, peeled
2 garlic cloves, crushed
½ tsp salt
½ tsp pepper
½ tsp cumin
750ml tomato juice
200g cucumber, diced
200g onion, diced
70g green pepper, diced

Liquidise the canned tomatoes, stock, oil and vinegar. Cut remaining tomatoes into small pieces. Mix garlic, cumin, salt and pepper. Add tomatoes, tomato juice and chopped vegetables and chill.

258 Goulash soup

30g vegetable oil
150g onion, chopped
227g stewing beef, cubed
150g tomatoes, canned
70g green pepper, chopped
1000ml beef stock
1 garlic clove, crushed
1g caraway seeds, crushed
1 tsp paprika
1 tsp salt
227g potatoes, peeled, diced

Heat oil in saucepan and fry onion. Add the meat and brown lightly on all sides. Add remaining ingredients except for the potatoes. Cover and simmer for 1 hour. Add potatoes and simmer for 25 minutes.

Weight loss: 56%

264 Lentil soup

60g butter
150g onion, chopped
2 garlic cloves, crushed
15g coriander
2g cinnamon
2g chilli powder
250g red lentils
1500ml water
30g tomato purée
1g salt

Heat butter in a saucepan and fry onion and garlic. Add spices and cook for 2 minutes. Add lentils, water, tomato purée and salt. Cover and simmer for 30 minutes.

Weight loss: 44%

267 Minestrone soup

50g bacon, chopped
50g onion, chopped
1 garlic, crushed
25g parsley, chopped
25g margarine
100g cabbage, chopped
80g carrot, chopped
80g leek, chopped
80g potato, chopped
50g runner beans
30g celery, chopped
10g tomato purée
1000ml vegetable stock
100g macaroni
½ tsp salt
½ tsp pepper

Fry chopped bacon, onion, garlic and parsley in margarine. Add remaining diced vegetable, tomato purée and stock. Cook for 20 minutes. Add pasta, seasoning and simmer for a further 15 minutes.

Weight loss: 12%

271 Mulligatawny soup

50g onion, chopped
125g carrot, chopped
125g swede, chopped
50g apple, peeled, cored, chopped
50g bacon, chopped
50g butter
25g flour
15g curry powder
15g tomato purée
25g mango chutney
1000ml beef stock
pinch of cloves
pinch of mace
50g rice
45ml double cream

Fry onion, carrot, swede, apple and bacon in butter. Add flour, curry powder, tomato purée and chutney and cook gently for a few minutes, stirring all the time. Add stock and spices and bring to the boil. Simmer for 30 minutes. Sieve or liquidise and return to the pan with the rice. Boil for 12 minutes. Stir in cream.

Weight loss: 19%

275 **Pea and ham soup**

454g dried green split peas
1250ml vegetable stock
1250ml water
100g ham, canned, chopped
60g celery, chopped
½ tsp tarragon
30g parsley, chopped
60g butter
250g carrots, chopped
250g onion, chopped
150g leek, chopped
1g pepper

Rinse peas and combine with stock and water in a pan. Bring to the boil. Add ham, celery, tarragon and 10g of the parsley. Reduce heat, simmer for 45 minutes. Melt butter and fry carrots, onion and leek. Cook for 10 minutes and add to stock. Simmer for 30 minutes. Remove ham, shred meat, return meat to soup. Add pepper and remaining parsley.

Weight loss: 17%

276 **Potato and leek soup**

30g butter
300g leeks, chopped
300g potatoes, chopped
750ml vegetable stock
2g salt
1g pepper
20g flour
250ml milk

Melt butter in a pan, add leeks, potatoes, stock and seasoning and bring to the boil. Simmer for 30 minutes. Add flour blended to a smooth cream with the milk. Simmer for 5 minutes.

Weight loss: 14%

277 **Scotch broth**

900g lamb, chopped
2000ml water
75g pearl barley
350g turnip, chopped
200g carrot, chopped
200g onion, chopped
200g leek, chopped
100g cabbage, shredded
125g peas
½ tsp salt
½ tsp pepper
30g parsley, chopped

Put the meat into a saucepan, add the water and bring to the boil. Add barley. Simmer for 30 minutes, then add turnip, carrots, onion and leek and simmer for 1½ hours. Add cabbage and peas and simmer for 20 minutes, adjust seasoning and add parsley.

Weight loss: 19%

283 **Vegetable soup**

100g carrots, chopped
100g onions, chopped
60g leeks, chopped
60g potatoes, chopped
60g turnip, chopped
50g celery, chopped
50g butter
700ml vegetable stock
50g peas
½ tsp salt
½ tsp pepper

Fry chopped vegetables in butter till soft. Add stock, peas and seasoning and simmer for 30 minutes.

Weight loss: 12%

288 Apple sauce, homemade

400g apple, peeled, cored, chopped
120ml water
60g sugar
10ml lemon juice

Put apple to a heavy based saucepan. Stew in water for 10 minutes. Add sugar and lemon juice and stir.

Weight loss: 17%

290 Barbecue sauce, homemade

50g butter
150g onion, chopped
5g tomato purée
30g Demerara sugar
30g Worcestershire sauce
30ml vinegar
10g mustard powder

Melt butter in saucepan and fry onion. Stir in tomato purée. Blend together remaining ingredients and stir into onion mixture. Simmer for 10 minutes.

Weight loss: 31%

303 Dressing, French, homemade

75ml olive oil
25ml vinegar
½ tsp salt
½ tsp pepper

Shake the ingredients together in a screw-topped jar or bottle.

305 Dressing, oil and lemon

75ml vegetable oil
25ml lemon juice
3g Demerara sugar
½ tsp salt
½ tsp pepper

Shake the ingredients together in a screw-topped jar or bottle.

309 Dressing, yogurt, homemade

125g yogurt
2 tsp lemon juice
3g mint, chopped
½ tsp salt
½ tsp pepper

Mix all ingredients together.

312 Guacamole

195g avocado flesh
20ml lemon juice
85g tomato, chopped
1g salt

Mash avocado and sprinkle with lemon juice. Add finely chopped tomato and salt.

313 Hollandaise sauce, homemade

5ml lemon juice
5ml vinegar
½ tsp white peppercorns
2 bay leaves
4 egg yolks
225g butter
½ tsp salt
½ tsp pepper

Put lemon juice, vinegar, peppercorns and bay leaf in a saucepan and bring to the boil. Reduce liquid by half and cool. Strain reduced vinegar, add egg yolks and whisk over a gentle heat until mixture thickens. Add butter and seasoning.

Weight loss: 14%

317 **Mayonnaise, homemade**

2 egg yolks
2g mustard
½ tsp salt
½ tsp pepper
2 tsp lemon juice
250ml olive oil

Whisk egg yolks, mustard, salt and pepper and 1 tsp of lemon juice to combine. Add oil gradually, whisking constantly. Add remaining lemon juice.

320 **Mint sauce, homemade**

20g mint leaves, chopped
20g sugar
30ml boiling water
45ml vinegar
2g salt

Chop mint leaves with 1 tsp of sugar. Place in bowl and add boiling water. Stir in remaining sugar, vinegar and salt.

322 **Pasta sauce, white**

25g butter
1 garlic clove, chopped
500g mushrooms, sliced
250ml single cream
125g ham, diced
90g Parmesan cheese

Heat butter in a saucepan. Add garlic and mushrooms and cook for 5 minutes. Heat cream in a double saucepan, add mushroom mixture, ham, Parmesan cheese and heat through.

Weight loss: 5%

324 **Raita**

100g cucumber
225g yogurt
3g mint leaves, chopped
2g cumin
1g chilli powder
½ tsp salt
½ tsp pepper

Grate or thinly slice the cucumber, mix with remaining ingredients.

333 **Sauce, tomato base, homemade**

440g canned tomatoes
20ml olive oil
70g onion, chopped
1 garlic clove, crushed
20g tomato purée
½ tsp sugar
5g basil, chopped

Chop tomatoes. Heat oil, add onions and garlic and cook until soft. Add remaining ingredients, cover and simmer gently for 30 minutes.

Weight loss: 18%

341 **Chutney, apple, homemade**

500g cooking apples, peeled, cored, chopped
400g onion, chopped
100g raisins
400ml vinegar
1 tsp salt
½ tsp pepper
1 tsp curry powder
½ tsp ground ginger
½ tsp mustard
450g granulated sugar

Mix all the ingredients except the sugar and boil gently until soft. Add the sugar and boil for a further 30 minutes.

Weight loss: 23%

346 Chutney, tomato, homemade

1000g tomatoes, peeled, chopped
125g cooking apples, peeled, cored, chopped
500g onions, chopped
100g sultanas
450ml vinegar
½ tsp mustard
2 tsp curry powder
1 tsp salt
½ tsp pepper
500g granulated sugar

Mix all the ingredients except the sugar and boil gently until soft. Add the sugar and boil for a further 30 minutes.

Weight loss: 38%

373 Stuffing, sage and onion, homemade

224g onion, sliced
112g white breadcrumbs
4g fresh sage, chopped
56g margarine
1 egg
¼ tsp salt
¼ tsp pepper

Parboil onions, drain and chop, mix with breadcrumbs, add sage. Melt margarine and add to stuffing. Mix thoroughly. Stir in egg and seasoning.

Weight loss: 19%

REFERENCES TO TABLES

1. Cutrufelli, R. and Matthews, R. H. (1986) *Composition of foods: beverages, raw, processed and prepared.* Agriculture Handbook No. 8-14, US Department of Agriculture, Washington DC

2. Cutrufelli, R. and Pehrsson, P. R. (1991) *Composition of foods: snacks and sweets, raw, processed and prepared.* Agriculture Handbook No. 8-19, US Department of Agriculture, Washington DC

3. Lewis, J. and English, R. (1990) *Composition of foods, Australia.* Volume 5, nuts and legumes, beverages, miscellaneous foods. Department of Community Services and Health, Canberra.

4. Marsh, A.C. (1980) *Composition of foods: soups, sauce and gravies, raw, processed and prepared.* Agriculture Handbook No 8-6, US Department of Agriculture, Washington DC

FOOD INDEX

Foods are indexed by their food number and **not** by their page number, except for those food ingredients and additives whose proximates and energy value have been given for labelling purposes in the Appendix on page 170.

This index includes three kinds of cross-reference. The *first* is the normal coverage of alternative names (e.g. Alcohol-free lager see **Lager, alcohol-free**). The *second* is to common examples or components of generically described foods, including selected brand names, which although not part of the food name have in general been included in the product description (e.g. Anchor half-fat butter see **Blended spread (40% fat)**). And the *third* is for foods related to those in this book but whose nutritional value has already been covered in other supplements and are not repeated here (e.g. Baby milks see ***Milk Products and Eggs* suppl.**). Full references to these supplements are given on page 5.